M. IRVING

FLOOR MAINTENANCE MATERIALS

THEIR CHOICE AND USES

FLOOR MAINTENANCE MATERIALS

THEIR CHOICE AND USES

J. K. P. EDWARDS, T.D., M.A.
Technical Director
Russell Kirby Limited

LONDON
BUTTERWORTHS

THE BUTTERWORTH GROUP

ENGLAND
Butterworth & Co (Publishers) Ltd
London: 88 Kingsway, WC2B 6AB

AUSTRALIA
Butterworths Pty Ltd
Sydney: 586 Pacific Highway, NSW 2067
Melbourne: 343 Little Collins Street, 3000
Brisbane: 240 Queen Street, 4000

CANADA
Butterworth & Co (Canada) Ltd
Toronto: 14 Curity Avenue, 374

NEW ZEALAND
Butterworths of New Zealand Ltd
Wellington: 26–28 Waring Taylor Street, 1

SOUTH AFRICA
Butterworth & Co (South Africa) (Pty) Ltd
Durban: 152–154 Gale Street

First published in 1969
Second impression 1972
Third impression 1974

© J K P Edwards 1969

ISBN 0 408 48690 2

Printed in England by
Fletcher & Son Ltd, Norwich

FOREWORD

In my capacity as Managing Director of a general cleaning company and National Chairman of the British Institute of Cleaning Science, I know only too well the costly pitfalls into which the unwary can fall—and no wonder: prior to this book there has never been such a comprehensive, factual and informative work to which one can refer. The text is based on many years of sound, 'down-to-earth' experience. It contains many useful practical tips and should be at the elbow of every person responsible for floor maintenance.

The author is a leading authority on the subject of floor maintenance. He has lectured extensively and written many articles on various aspects of cleaning. After graduating from Cambridge, his first introduction to industry was in the paint trade. He was subsequently appointed Technical Director of a company manufacturing floor maintenance materials and has been awarded a Fellowship of the British Institute of Cleaning Science for his services to education.

This book will fill a long-felt want. I compliment the author on the painstaking research and preparation which has gone into compiling the material and on his initiative in writing this, which is, to the best of my knowledge, the first book of its kind.

J. B. KING, F.B.I.C.Sc., L.M.R.S.H.
National Chairman, British Institute of Cleaning Science

PREFACE

It has been said that the cost of maintaining a building over a period of twenty-five years is approximately the same as the initial cost of the building itself. The cost of floor maintenance is a large proportion of the total maintenance costs.

Floors are a big investment. Approximately 80% of dirt entering a building is carried in on footwear. If floors are not properly maintained they may suffer damage and need to be replaced long before it would otherwise be necessary, thereby incurring considerable, unnecessary, expense.

Perhaps the first and lasting impression that a visitor, or employee, receives on entering a building is the state of the floor. An attractive floor, well maintained, gives an immediately favourable impression and boosts morale. On the other hand, a floor maltreated or ruined by neglect reflects badly on the owners. In old buildings, a clean, well maintained floor can do much to elevate the overall tone and to brighten the working conditions of all.

Well maintained floors are clean floors. Correct maintenance not only keeps floors clean but also ensures they are safe and hygienic. Not only are floors made more attractive but by protecting them from dirt and damage they last longer and reduce maintenance costs.

During recent years floor maintenance technology has advanced at a rapid rate. Materials unheard of a few years ago are now commonplace. Those responsible for maintenance are liable to become a little confused regarding the maintenance of the many and varied types of floors.

The aim of this book is to explain, in everyday, layman's language, the many types of materials available for the maintenance of floors. The advantages and disadvantages of each material are discussed fully. The book is intended as a practical guide rather than a theoretical work. A section has been allocated to ' Faults—Causes and Remedies ', compiled from experience in the industry gained over a period of many years.

A ' Glossary of Technical Terms ' is appended to describe, again in layman's language, terms frequently found in journals, articles and trade literature.

PREFACE

Discrimination in selection of materials should be the policy of all engaged in the maintenance of floors. It is hoped that this book will provide the reader with sufficient information to enable the correct materials to be used, with consequent benefit to all concerned.

J. K. P. EDWARDS

ACKNOWLEDGEMENTS

I AM most grateful to all those whose kind help have made this book possible. In particular I should like to thank Mr. L. Harold Russell, J.P., Chairman and Managing Director of Russell Kirby Ltd., for his encouragement and constructive advice. I also much appreciate the assistance given by Mr. J. H. Arnison of Riversdale Technical College, and my grateful thanks are due to Mr. B. G. Taylor, Technical Manager of Russell Kirby Ltd., for his practical assistance.

Certain sections of this book have been adapted from articles by myself that have appeared in the following journals: *Cleaning and Maintenance*; *Shop Equipment and Shopfitting News*; *Caterer and Hotelkeeper* and *Flooring*.

I should like to express my thanks to the Editors of these journals for their permission to use this material.

Last, but by no means least, I am very grateful to my wife, Lesley, and to Mrs. B. McArthur for their interest and care in the typing of this book.

CONTENTS

Foreword v
Preface vii
Acknowledgements ix

1 Soaps and Industrial Detergents 1
2 Floor Seals 26
3 Preparation of Floors for Sealing and Application of Seal 88
4 Floor Waxes 106

Appendix I. Floor Maintenance Chart 151
Appendix II. Faults—Causes and Remedies 155
Appendix III. Glossary of Technical Terms 169
Appendix IV. Conversion Tables 179
Index 183

1

SOAPS AND INDUSTRIAL DETERGENTS

INTRODUCTION

DIRT has been defined as 'matter in the wrong place'. Removal of dirt, that is, cleaning, is a constant and growing problem in our modern civilization.

In hospitals, schools, factories and offices the presence of dirt can be a problem of great importance. In some areas, for example operating theatres and computer rooms, the presence of dirt can be critical.

There can be no simple answer to all the many and varied cleaning tasks that arise. Much research has been carried out to determine how best to overcome each obstacle and research chemists are constantly striving to improve materials and methods.

The purpose of maintenance cleaning is industrial housekeeping. Its object is not only to remove waste materials which might become health, safety or fire hazards, but also to maintain buildings in a clean and attractive condition, for the benefit of all who use them.

John Wesley placed cleanliness next to godliness. In more recent times the passing of the Public Health Acts towards the end of the last century reflected the growing awareness of the need for improvements in public hygiene. This has been emphasized by the Clean Air Act and by the Offices, Shops and Railway Premises Act of 1963.

The labour cost in any maintenance cleaning operation has been found to be approximately 90% of the total cost. Selection of the correct cleaning material is, therefore, essential to reduce labour cost to a minimum. The success of any cleaning operation depends upon the selection of the correct materials for the task in hand. Industrial cleaning is a complex operation that calls for a considerable degree of technical knowledge and practical 'know-how'.

To understand cleaning, those responsible should possess a knowledge of the nature of dirt and grime to be removed, the properties of the materials to be used, of the surfaces from which dirt is to be removed and, equally important, a working knowledge of the correct procedures and methods to be employed.

HISTORY

Soap has an extremely long history. Many textbooks quote the passage in the Old Testament that can be found in Jeremiah, 'For though thou wash thee with nitre and take thee much sope', as showing that soap was in use in those far off days. It was, however, more likely to have been a form of soda similar to that occurring naturally in Egypt, or some form of plant ash used as a primitive cleanser.

The first mention of soap in Britain was made in A.D.1000. Even by the end of the sixteenth century soap was considered a luxury. It was said of Queen Elizabeth I that she 'hath a bath every three months whether she needeth it or no'.

Consumption of soap in Great Britain in 1861 was about 100,000 tons a year. Today consumption has risen to about 600,000 tons a year, or over 24 lb/head of the population.

Soap was, and still is, one of the best and cheapest detergents. It has, however, several limitations which prohibit its use under all conditions; for example, in hard water soap forms a 'scum' which cannot easily be removed.

Towards the end of the last century work was carried out both in Europe and America to find an alternative to soap that was not based on natural oils and fats. This was because vegetable oils and fats were needed for foodstuffs and the demand for soap was increasing. In 1913 a Belgian chemist, Reychler, manufactured in the laboratory a number of pure surface-active agents, or 'surfactants' as they are called, whose properties were very similar to those of soap. The most important of these was sodium cetyl sulphonate, $C_{16}H_{33}SO_3Na$.

During World War I the shortage of natural oils in Germany became acute and considerable development work on the production of soapless detergents was carried out. In 1916 a German firm took out a patent for an alkyl sulphonate sold commercially under the trade name of Nekal A.

Between the wars there was a period of intense research and development, which started in Britain and Germany and spread to the U.S.A. During World War II the rapidly increasing need for detergents stepped up both production and research into new materials and methods.

Today there are thousands of surfactants, possibly tens of thousands, which have, to some degree, the properties of soap. It

REQUIREMENTS OF A DETERGENT

is estimated that about 350,000 tons of liquid and solid synthetic detergents are used in Britain annually.

DEFINITIONS

The Oxford dictionary defines a detergent as 'something which cleans'. The word 'detergent' originates from the Latin, *de*—off and *tergere*—to wipe or clean. Water, therefore, is technically a detergent, as also is soap.

Water can be used as the sole agent to loosen and remove dirt if there is sufficient agitation. Even today, in primitive societies, washing is carried out on the banks of a stream or pool, beating and rubbing soiled linen for hours to remove the dirt. The linen is eventually cleaned, but the time required is very considerable and the life of the linen is shortened by the washing action.

Soap reduces the surface tension of water and is, therefore, known as a 'surface-active agent', or 'surfactant'. It has the property both of dissolving in water and of dispersing grease and thus is widely used for washing purposes.

NATURE OF DIRT

Dirt consists of a wide variety of different materials. Particulate dirt, consisting of fine particles of soot, rust, clay, soil and many other items, adheres very tenaciously to surfaces; in general, the smaller the particle size, the harder it sticks. Most grit and dust will be of a siliceous nature, generally in the form of oxides.

In most industrial conurbations, dirt is inevitably associated with grease from petrol, lubrication and diesel oil fumes, together with condensation products from chimneys, flues and exhausts. Frequently, specks of particulate dirt are embedded in a film of grease.

Organic matter in soilage is to be found everywhere, and can consist of animal or vegetable debris or bacteria-laden soil.

REQUIREMENTS OF A DETERGENT

An industrial detergent, to be acceptable, must satisfy a number of requirements. It must be readily soluble in water so that it can be used in an aqueous medium; it must 'wet' the surface to which it is applied; loosen and remove the dirt; emulsify and disperse the dirt in the liquid; prevent it from being re-deposited and be easy to rinse, when required.

Household detergents must have many additional requirements, but these are not considered here because they are not relevant to industrial materials.

Action as a Wetting Agent

Water, by itself, is not a good wetting agent. If, for example, a dry plate or cup is rinsed in water, removed and examined, the surface will be found to be unevenly wetted, the water having shrunk back from some areas. If, however, a detergent is added to the water a far greater area will be covered. This is because the surface tension has been reduced, allowing the water to spread over, or wet the surface. Cleaning cannot begin until the washing liquid has come into intimate contact with both the surface to be cleaned and the dirt itself. The presence of detergent lowers the surface tension and promotes contact between the cleaning liquid, surface and dirt. This can be illustrated, diagrammatically, as in *Figure 1.1*.

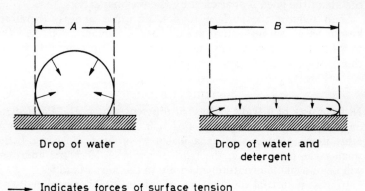

Figure 1.1. The effect of detergent on surface tension: (A) area covered by drop of water, (B) greater area covered by drop of water to which detergent has been added

Removal of Dirt

Once the surface and dirt are thoroughly wetted the detergent must be able to fasten on to the dirt and lift it from the surface being cleaned. Detergents are formulated so that each molecule consists of two parts, a 'head' and a 'tail'. The head is the hydrophilic, or water-loving, component and the tail the hydrophobic, or water-hating, component.

The hydrophobic components penetrate and stick to the dirt, grease or oil being removed and the hydrophilic components remain in the water. The result is that the dirt, under physical agitation,

REQUIREMENTS OF A DETERGENT

is lifted from the surface and held in the liquid in a thin coating of detergent. *Figure 1.2* illustrates diagrammatically the removal of dirt.

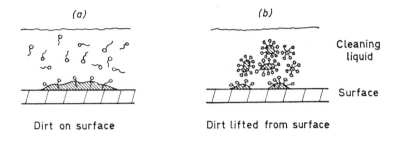

Dirt on surface   Dirt lifted from surface

∽ Detergent molecule. The round 'head' represents the hydrophilic (water-loving) component and the long 'tail' the hydrophobic (water-hating) component

Figure 1.2. The removal of dirt by detergent

Emulsification and Suspension of Dirt

Once the dirt is in the cleaning liquid it must be broken down into small particles, dispersed and held in suspension to prevent it from being re-deposited elsewhere. The detergent itself is an emulsifier and will emulsify the dirt, grease or oil, in water. Detergents are formulated so that once dirt is in suspension it is prevented from coagulating into larger particles and so remains dispersed. This is because minute electrical charges, transferred to the dirt particles from the detergent molecules, cause the dirt to become charged with the same polarity, thereby making them repel one another.

Ease of Rinsing

Dirt must be removed by rinsing once it is dislodged from the surface being cleaned. This can be achieved by many different methods, depending on the item or surface. The detergent must, however, hold and disperse the dirt sufficiently to enable all dirt to be wiped or rinsed away so as to leave a clean, dirt-free surface. A detergent must not only clean the surface, but also leave it completely unharmed in any way.

SOAPS AND INDUSTRIAL DETERGENTS

Before each type of detergent is discussed in detail, mention must be made of a method of measuring the acidity or alkalinity of a detergent, by means of the pH scale.

pH SCALE

By using pH, or potential of hydrogen, acidity and alkalinity can be expressed in simple numerical terms. The pH scale ranges from 0 to 14. pH 0·1 is the value given by a Normal solution (N) of hydrochloric acid in water (i.e. 36·5 g hydrochloric acid/1000 ml water). At the other end of the scale, pH 14 is the value given by a Normal solution of caustic soda in water (i.e. 40 g caustic soda/1000 ml water).

The pH scale is only used in connection with solutions which are near neutrality. If the acid or alkali is stronger than those shown above, and many are, then the scale will not apply.

A pH of 7·0 is neutral and is the value given to pure distilled water. Materials with a pH value below 7 are acidic, the acidity increasing as the pH value decreases. Materials with a pH value above 7 are alkaline, the alkalinity increasing as the pH value increases (*Figure 1.3*).

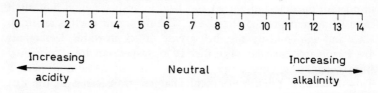

Figure 1.3. pH Scale

Each whole number increase indicates an increase in alkalinity in magnitude of 10. A pH of 9·0, for example is ten times more alkaline than 8·0; a pH of 10·0 is ten times more alkaline than 9·0 and one hundred times more alkaline than 8·0, etc.

It must be mentioned that a certain amount of alkalinity or, in some cases, acidity, is not only desirable but necessary for effective cleaning.

The pH values of some common materials are given in *Table 1.1*.

IMPORTANCE OF USING THE CORRECT DETERGENT

Table 1.1. pH Values

Caustic soda	N	14·0
Caustic soda	0·1 N	13·0
Caustic soda	0·01 N	12·0
Ammonia	N	11·6
Ammonia	0·1 N	11·1
Ammonia	0·01 N	10·6
Household soap in solution		8·8 (approx.)
Toilet soap in solution		9·0 (approx.)
Normal blood		7·4
Tears		7·2
Perspiration		4·5
Gastric Juices		1–3
Hydrochloric acid	0·01 N	2·0
Hydrochloric acid	0·1 N	1·0
Hydrochloric acid	N	0·1

IMPORTANCE OF USING THE CORRECT DETERGENT

Detergents are formulated to meet specific needs. Whereas the correct material will undoubtedly carry out its cleaning task effectively and well, the wrong material may, at best, be ineffective and at worst ruinous to the floor. Numerous examples are to be found of detergents being used wrongly.

The main corridors of a large secondary school suffered from the effects of a build-up of water emulsion floor wax, accumulated over a period of several years. A yellow-brown build-up of wax and dirt was evident over the whole floor and particularly along the edges of the corridors.

The caretaker reported that the floor wax was extremely difficult to remove, even when scrubbed with a nylon web pad under an electric polishing/scrubbing machine. On investigation it was found that a neutral detergent was being used for routine cleaning and in attempts to strip the build-up of floor wax. Once a properly formulated alkaline detergent was used this build-up was stripped from the floor easily. A maintenance system, including periodic stripping of wax using an alkaline detergent, was then put into operation.

Terrazzo floors are sometimes found to be dirty and slippery due to treatment with soap. Soap fills the open pores in the terrazzo and leaves a slippery film on the surface; dirt adheres to the soap and is often difficult to remove. By use of the correct detergent the appearance can be rapidly improved and the slippery condition overcome.

Several floors have been ruined in recent years by using caustic type materials. A green linoleum floor, laid in the corridors of a hospital, turned brown after scrubbing with a solution of caustic soda in water. The caustic solution was far too strong and irreversibly changed the colours of the pigments in the linoleum.

Discrimination in the selection of a detergent for any cleaning operation is essential, therefore, before any cleaning task is started, the correct type of detergent for that task should be ascertained and if possible, trials should be carried out before the detergent is used on a large scale. Trials are especially important if the quality of the flooring material to be cleaned is at all suspect. Some, particularly very cheap floor coverings, are affected by alkaline material. A test on about 1 ft^2 of flooring, preferably in an unused corner, will quickly determine whether the coloured pigments are affected by the detergent. If pigments are discoloured a milder detergent must be used.

Tests are often omitted, sometimes with disastrous results. It is always advisable before starting to ensure that the desired results can be achieved using the selected detergent.

TYPES OF DETERGENT

No single detergent can be expected to cope with the many and varied cleaning tasks that arise in industrial surroundings. A range of detergents has been developed over the years, each designed to meet a specific requirement. There is, today, a growing demand for the more sophisticated cleaning agents, particularly in heavily industrialized areas.

The types of detergent that are discussed include:

      Soap
      Anionic detergents
      Non-ionic detergents
      Cationic detergents
      Amphoteric detergents
      Alkaline detergents
      Caustic materials
      Acid cleaners
      Detergent crystals

In addition, it is convenient to include under this heading a group of materials known as 'solvent-based detergent wax removers', used for removing solvent-based waxes, oil and grease.

Soap

Soap has been defined as ' the product formed by the saponification or neutralization of fats, oils, waxes, rosins, or their acids with organic or inorganic bases '. Descriptive adjectives are usually

TYPES OF DETERGENT

applied to the name 'soap', to indicate various properties. Examples are 'bar' soap, 'liquid' soap, 'soft' soap and 'salt-water' soap.

By further definition, soap is generally understood to mean any water-soluble salt of those fatty acids which contain eight or more carbon atoms.

Soap can be represented by the formula

$$R_1\text{—}COOR_2$$

where R_1 is a straight chain hydrocarbon radical with between eight and twenty-one carbon atoms and R_2 is a base-forming radical, such that R_1—$COOR_2$ is soluble in water.

A typical example of a soap is sodium stearate, the formula being:

$$CH_3(CH_2)_{16}COONa$$

Sodium stearate

Perhaps the most lasting and greatest virtue of soap is its very long record of safety to human beings. Since in many cases soaps are products derived from normal animal metabolism, this is to be expected. Its excellent biological properties also extend to the microbiological world—soap is one of the few surfactants that will not cause stream pollution.

Another advantage of soap is its effect upon bactericidal agents. Some bactericides, for example hexachlorophene, are most effective on the human skin when applied from a soap medium, a property which is of great importance in hospitals and the medical field. Considerably higher concentrations of these bactericides are needed to achieve the same effects when used in conjunction with synthetic detergents.

Soap has a slippery feel, will form gels, emulsify oil and lower the surface tension of water. It also has detergent and wetting properties and will give a foam when shaken in soft water. Unlike synthetic detergents, it does not need soil-suspending agents and does not need foam stabilizers.

The shortcomings of soap when used in hard water are well known. Calcium and magnesium salts present in hard water react with it to from an insoluble compound, or scum. Where natural lime or chalk is present in the water the hardness can become very marked.

Today, soap is not used to any great extent for cleaning industrial floors, having been displaced by the advent of synthetic detergents. One reason is that considerable rinsing is required to remove any scum left on the floor after using soap. It may also leave a greasy film which is likely to build up and cause slippery conditions.

Perhaps the most important reason, however, is its higher cost, particularly in hard-water areas. Sufficient soap must first be added to the water to neutralize the insoluble salts present before it can be made available for cleaning purposes.

Difference between Soap and Soapless Detergents

Whereas soaps are made from naturally occurring fats and oils, soapless or synthetic detergents (syndets), are manufactured from processed organic chemicals, usually derived from petroleum.

The molecular structure of soap and soapless detergents is basically the same, the difference depending mainly on the physical properties of the hydrophilic, or water-attracting component.

Examples of typical soap and soapless detergent molecules are as follow:

Typical soap molecule: $C_{17}H_{35} \cdot COONa$ (Sodium stearate)

Typical soapless
detergent molecule: $C_{12}H_{25} \cdot O \cdot SO_3Na$ (Sodium dodecyl sulphate)

In water, both molecules split into two, one part being hydrophobic and the other hydrophilic.

	Hydrophobic component	Hydrophilic component
Soap molecule	$C_{17}H_{35}$	COONa
Detergent molecule	$C_{12}H_{25}$	$O \cdot SO_3Na$

The difference between the two hydrophilic components accounts for the difference in behaviour between soap and soapless detergents in hard water. Whilst the calcium and magnesium salts present in hard water form an insoluble scum with soap, in combination with soapless detergents a water-soluble compound is formed which does not precipitate or form a scum in hard water. This gives the soapless detergents a great advantage, particularly in hard water areas.

Another advantage of soapless detergents over soap is that many can be used satisfactorily in an acid medium and still retain their detergent properties. This gives them desirable degreasing properties. Grease, as opposed to dirt, is soluble in detergent because of the properties of the hydrophobic component, which is attracted to grease in preference to water. The detergent breaks the grease down into minute droplets so that it can be rinsed away. In addition, soapless detergents do not leave a residual film after use but dry, leaving the surface clean and bright.

TYPES OF DETERGENT

Anionic Detergents

Composition

An anionic, commonly known as a ' neutral ' detergent, is one in which the active part of the detergent molecule carries a negative charge when in solution in water. The main constituents of anionic detergents include alkyl aryl sulphonate, alkyl sulphates and sulphonates, sulphated and sulphonated amides and amines, sulphated and sulphonated esters and ethers and many other types.

By far the most important are the alkyl aryl sulphonates, comprising almost the whole of the washing powders manufactured today.

Basically, anionic detergents are manufactured from strong alkalis and weak acids. Because of this they are alkaline in nature, possessing a pH value of over 7 and generally between 7 and about 9. A small proportion of non-ionic detergent is normally included, in addition to other additives, to assist in foam producing and detergency powers. It has been found that a mixture of an anionic and non-ionic detergent behaves synergistically, that is, it is superior in performance to either of the individual components acting alone.

Uses

Anionic detergents are by far the most widely used, and today form the basis of over 80% of all soapless detergents used on the domestic and industrial markets. Powder materials are normally of the spray-dried type and resemble small beads in apppearance.

Liquid anionic detergents are typified by the washing-up type of detergent, usually pale straw in colour, and are free-flowing liquids that disperse easily in water. In floor maintenance the liquid form is generally preferred because of its greater convenience. When using powder detergents time must be allowed for the detergent to dissolve.

Anionic detergents should be added to water, either warm or cold, in accordance with the manufacturer's instructions. The amount to be added will depend on the active content, or ' strength ', of the detergent and on the cleaning task to be carried out. The solution can be used with absolute safety on either unwaxed floors or floors treated with a water emulsion floor wax or solvent-based wax. The detergent will soften and remove light dirt, soilage and carbon black marks. Anionic detergent solutions can be used in conjunction with either mopping equipment or a polishing/scrubbing machine. As anionic detergents produce foam, sometimes in large quantities, only the recommended amount should be used and an excess should be avoided.

Anionic detergents are safe for use on all flooors and will not cause floors to lift or affect pigment colourings in floor covering materials. Excessive use of water on any floor should, however, be avoided, as water itself, if allowed to remain too long in contact with the floor, may cause some harm. Some adhesives used to secure tiles to the sub-floor are water soluble. If water is allowed to penetrate between the tiles it could weaken the adhesive and in extreme cases cause tiles to lift.

Water should also be kept to a minimum on wood, wood composition, cork and magnesite floors, even if the floors are sealed, as water tends to remove the colour from wood and causes splintering.

After using an anionic detergent the floor should be rinsed and allowed to dry.

Anionic detergents can safely be used on floors treated with a water emulsion floor wax and will remove dirt without harm to the wax. This is because the pH value of an anionic detergent is in the region of neutrality and emulsion floor waxes, in the main, are unaffected by neutral detergent solutions. Anionic detergents will not remove solvent-based waxes either, as solvent and water do not mix. In both cases, having rinsed the floor it should then be buffed as required, using a polishing machine in conjunction with a fine grade metal fibre or nylon web pad.

Non-ionic Detergents

Composition

Non-ionic detergents differ from anionic detergents in that when added to water the detergent molecules dissolve as intact molecules, that is, they do not ionize or carry a charge. The main constituents of non-ionic detergents include condensates of ethylene oxide and the esters or ethers derived from poly-alcohols.

Basically, non-ionic detergents are manufactured from alkalis and acids of equal strengths. They have, therefore, a pH value of 7 and are neither alkaline nor acid.

The growth of non-ionics in recent years has been very considerable and second only to that of the alkyl aryl sulphonates, or anionic detergents. This is because non-ionic detergents can easily be included in a wide variety of formulations, as they are compatible with many other ingredients. When used on their own, however, they have poor foaming powers and poor foam stability.

TYPES OF DETERGENT

Non-ionic detergents are generally higher in cost than anionics. They are largely available as liquids and therefore can only be used in very limited amounts in powder or solid detergents.

Uses

One of the main uses of non-ionic detergents is to act as a foam booster in conjunction with other detergents, generally of the anionic type. As stated above, a mixture of anionic and non-ionic detergent behaves synergistically, being superior in performance to either of the individual components acting by itself.

When used for general purpose industrial cleaning, non-ionic detergents are generally more efficient as cleaners than are anionic detergents. This is because they are affected even less by hard water and are particularly efficient for removing grease. Their low foaming properties can be of advantage in industrial applications, making them very suitable for use in conjunction with scrubbing machines and other industrial cleaning equipment. It must be emphasized that there is little, if any, relationship between foam and cleaning ability. Excess foam can be a considerable nuisance, requiring many rinses to remove it completely.

As with anionic detergents, non-ionics should be added to water, either warm or cold, in accordance with the manufacturer's instructions. The solution can be used with absolute safety as non-ionics are safe for all surfaces. Absence of foam must not be taken as indicating an absence of cleaning power. The strength of the detergent, as with an anionic detergent, depends on the type and quantity of active material present, rather than on foam.

Once the floor is clean it should be rinsed and allowed to dry.

Table 1.2 shows comparative properties of non-ionic and anionic detergents.

Table 1.2. Comparative Properties of Non-ionic and Anionic Detergents*

Non-ionics	*Anionics*
Less foam—more easily controlled	Always considerable foam
Easier to rinse	More difficult to rinse
Less effective wetting of metal surfaces	Better wetting of metal surfaces
Less effective dirt dispersing power—needs added dispersant for heavy dirt loads	Greater dirt dispersing power—less redeposition of dirt, greater dirt carrying capacity
More effective in hard water	Not as effective in hard water
Very effective for removing oils and grease	Very effective for removing inorganic dirt and soil

* *After* J. B. Davidson, *Soap and Chemical Specialties* 1, (1959), 47.

Cationic Detergents
Composition

A cationic detergent is one in which the active part of the detergent molecule carries a positive charge when in solution in water. The main compounds comprising cationic detergents are the ammonium halides, or quaternary ammonium compounds. Perhaps the best known is benzalkonium chloride, which in the U.S.A. is marketed under at least twenty different brand names.

Cationic detergents are basically manufactured from weak alkalis and strong acids, and are therefore acidic in nature, possessing pH values of below 7.

Many quaternary ammonium compounds have gained in importance in recent years and now have a wide application in industrial and hospital fields as bactericides.

Uses

Used alone, cationic detergents possess little, if any, cleaning value, and for this reason are not used as detergents. They are generally combined with non-ionic detergents which give added detergency and foam and in combination together produce a satisfactory material. Cationic detergents cannot be blended with anionic detergents as they are rendered ineffective by them, being precipitated as electro-neutral salts.

Cationic materials possess valuable anti-static properties. They carry a positive charge in solution and transfer this to the surface being treated. Dust in the atmosphere also carries a positive charge. Because the surface treated with a cationic detergent carries a positive charge, dust is repelled by it. As a result, surfaces remain cleaner for a longer period of time.

By far the most important use of cationic materials is in the field of bactericidal detergents. Quaternary ammonium compounds, or ' quats ' as they are known, have excellent germicidal properties. They have been developed as sanitizing compounds, rather than as detergents, and are extensively used in hospitals and in similar applications for the disinfection of floors, walls, furniture and fittings. Many have an almost complete absence of odour, low skin irritation and high bacteriological activity. They are also finding increasing application as dish-washing detergents, particularly where communal feeding takes place, for example in canteens and restaurants.

At present there appears to be no really satisfactory explanation as to how the quaternary ammonium compounds attack bacteria.

TYPES OF DETERGENT

It appears that in any particular case the effectiveness of the 'quat' depends on the type of bacteria and the amount of 'quat' present.

Amphoteric Detergents

Composition

Conventional detergents are normally classified as anionic, cationic or non-ionic, depending upon the charge on the active portion in aqueous solution.

The amphoteric, or 'ampholitic' detergents as they are sometimes called, cannot be included in any of these types, as they have both acidic and alkaline properties. They are characterized by having an anionic and a cationic group in the same molecule.

Perhaps the most important property of amphoteric detergents is the effect which pH has on performance. Amphoterics, unlike most detergents, are greatly affected by changes in pH. Wetting powers, detergency, amount of foam and solubility are some of the properties affected by small variations in pH.

Their behaviour can be shown as follows:

Acid pH	*Neutral pH*	*Alkaline pH*
Cationic	Non-ionic	Anionic

At pH values of 8 and above, amphoteric detergents behave similarly to anionic detergents. In the neutral range, from pH 8 to 6 they are balanced and behave as non-ionics. At about pH 4 amphoterics are essentially insoluble. Below pH 4 they become soluble again and are cationic in behaviour.

At high pH values, therefore, amphoterics have excellent detergency powers, as good as conventional anionic types. At lower pH values, however, detergency powers are much reduced.

Uses

Commercial availability of amphoteric detergents has only been made possible in recent years. Amphoterics are used mainly in speciality detergent formulations. They are not used in large amounts as they are relatively expensive and for this reason cannot compete with conventional anionic detergents.

Their main outlets are in speciality hard surface cleaners, for example metal cleaners, steam cleaners and some oven cleaners, and are widely used in the industrial field, for example in the electroplating industry and in textile chemicals. Because they are non-toxic, non-irritating, germicidal and compatible with anionic, non-ionic and cationic detergents, they are also used in shampoos, medicated liquid soaps and aerosol shampoos, in limited quantities.

Alkaline Detergents

Composition

An alkaline detergent can be defined as a water-soluble alkali having detergent properties, but containing no soap. It is generally accepted that alkaline detergents range in pH value from about 9 to approximately 12·5.

Alkaline detergents are the work-horses of the cleaning industry, removing, as they do, a wider range of dirt and soil than any other type of detergent. They can be used in a wide variety of cleaning equipment and their solutions are very economical in use. Because of these advantages they are the most generally used of all cleaning materials.

The materials used in formulating alkaline detergents, both liquid and solid, include surface-active agents and a wide range of alkaline salts. The most commonly used are sodium carbonate, trisodium phosphate, sodium silicate and sodium tripolyphosphate. Sodium bicarbonate, sodium sulphate and certain silicates are also used to a lesser extent.

For maximum efficiency, high alkalinity is important to saponify fats and to neutralize acids found in many types of dirt. A number of the materials present in alkaline detergents are also very effective for removing oils and greases.

The total solids content alone is not necessarily indicative of the detergent. Strength depends mainly on the type of materials used, their relative proportions, and the pH value of the detergent in solution.

The better quality alkaline detergents exhibit relatively low foaming properties when compared with dish-washing, anionic detergents. Low foam is desirable because foam can be a nuisance when used for floor cleaning. Foam not only creates difficulties in machines used for applying a solution of detergent to the floor, but also necessitates several rinsings to ensure that all traces are removed from a cleaned floor. Rinsing is often a difficult and lengthy operation, particularly in buildings not fully equipped for cleaning purposes.

Uses

Alkaline detergents are used for application where a stronger type of detergent is required. They are used for 'hard surface' cleaning, removing water emulsion floor waxes from all types of floor (the latter probably being their main use), removing stubborn carbon black marks and heavy accumulations of dirt, cleaning very dirty walls and paintwork and many other tasks that require a relatively strong material.

Most modern water emulsion floor waxes are formulated in such a way that they are sensitive to alkaline detergents and can be removed by them.

Alkaline detergents must be powerful enough to carry out the tasks of cleaning very dirty surfaces and wax stripping, and because they are strong, care must be taken when using them on floor coverings, such as linoleum, which may be affected by them.

Over a period of time, strongly alkaline solutions will, by removing the linseed oil and affecting the wood flour, denature linoleum which is composed essentially of linseed oil, cork shavings, wood flour, colour pigments and other materials.

Under certain circumstances the coloured pigments can also be affected, with the result that the lino changes shade. Many of the cheaper grades of lino, containing pigments which are not stable to alkali, are particularly liable to be affected. This is especially so in the case of those containing green and blue colourings. Green pigments are made from blue and yellow pigments combined; blue is affected by alkali, with the result that it fades and loses colour. Blue floors, therefore, tend to lose colour and green floors assume a yellowish shade because the yellow is relatively unaffected. Thus, all shades of green and blue should be treated with caution. This also applies to the in-between shades, such as turquoise blue.

If the floor changes in shade, the degree of change will depend on the strength of the alkaline detergent used, the length of time it is allowed to remain in contact with the floor and the quality of the colour pigment used.

Care, therefore, should always be taken when using an alkaline detergent, particularly if it is to be used neat. If there is any doubt about the pigments in the floor covering, a test area should be treated to ensure the colourings are fast to alkali. This can be carried out by pouring a little neat detergent on to a cloth and rubbing a small area in one corner. If there is any evidence of colour change the test should be stopped immediately and a weaker solution used, until satisfactory results are obtained.

To strip emulsion floor wax, the recommended amount of liquid detergent is added to water and spread with a machine or mopping equipment over the area to be treated. It should then be allowed to remain on the floor for a few minutes to allow the detergent to

SOAPS AND INDUSTRIAL DETERGENTS

penetrate and loosen the floor wax. The floor is then scrubbed, using a polishing/scrubbing machine, in conjunction with a coarse grade metal fibre or nylon web pad, or deck scrub. Once the floor wax is loosened from the surface it can be removed by suction drier or mopping equipment. The floor must then be thoroughly rinsed to remove all traces of the detergent.

Rinsing is a particularly important operation and is often overlooked or carried out ineffectively. The aim of rinsing is to ensure that no alkaline residue remains on the floor. Most alkaline detergents consist of solid materials dissolved in water. If the floor is not rinsed, once the water has evaporated the solid material remains on the floor as a very fine white powder. This residue is alkaline by nature and as soon as emulsion floor wax is applied, the water present re-activates the alkaline residue forming an alkaline solution. This can attack the fresh application of emulsion floor wax and break it down as it dries. The result can be a loss of gloss and in extreme cases the emulsion floor wax breaks down into a fine white powder, so that the floor appears white.

When rinsing has not been carried out thoroughly an alkali residue may remain on the floor in patches. When a fresh application of emulsion floor wax is made, the result will be excellent in those areas rinsed thoroughly, with possibly some loss of gloss in the partly rinsed areas and, perhaps, whitening in the unrinsed areas. The overall result could be very patchy in appearance.

When rinsing it is always desirable to add a little vinegar to the rinsing water. Vinegar is mildly acidic and helps to neutralize any alkali remaining on the floor.

Once the floor has been thoroughly rinsed a new application of floor wax will give the best possible results.

It must, however, be recognized that it is not necessary for the floor surface to be chemically neutral before floor wax is applied, because emulsion floor waxes will tolerate some degree of alkalinity. Perhaps the most sensitive are the wash and wax types, comprising a wax and detergent in the one material. These require a surface as near to neutral as possible. As a general rule, however, emulsion floor waxes will give satisfactory results up to a floor pH value of about 8·5. If the pH value of the floor is between 8·5 and 9·5, substandard results can be expected, due to patchiness and loss of gloss. If the pH value is above 9·5, results will be poor.

Universal indicator papers can be used to measure the pH value of a floor. After rinsing, a paper is put onto the floor and the colour change evaluated on the scale provided with the papers. Accurate pH values cannot, of course, be obtained by this method, but values in the 8 to 9 range can be read satisfactorily with a little practice.

More recently, special materials have been developed for removing multi-coats of water emulsion floor wax, or 'build-up', which is often found along the edges of corridors. These special, or 'fortified' alkaline detergents, include a small amount of volatile solvents, which give extra power to the alkali and assist in breaking up the films of emulsion floor wax.

Most solvents are harmful to thermoplastic tiles, PVC(vinyl) asbestos, asphalt and similar surfaces. The solvents used in the 'fortified' alkaline detergents, however, evaporate as soon as they have carried out their task of breaking down the emulsion floor wax and do not harm the floor in any way. The loosened floor wax is then removed from the floor by using a scrubbing machine or deck scrub. When used in conjunction with abrasive nylon mesh discs, a fortified alkaline detergent will easily strip even the most difficult build-up of floor wax.

Alkaline powder detergents, of which the best are generally based on sodium tripolyphosphate or similar materials, are also generally formulated to have minimum foaming properties, so facilitating rinsing.

There are two main methods of using alkaline powder detergents. The first is to add the recommended amount to hot or boiling water. This solution is then poured into the tank of a scrubbing machine and dispensed during the scrubbing operation, or poured directly on to the floor and spread with a mop or applicator. It is allowed to remain on the floor for a few minutes to penetrate and loosen the dirt, and the floor is then scrubbed thoroughly with a scrubbing machine or deck scrub, rinsed well and allowed to dry.

The second method is particularly effective for cleaning very dirty floors. The floor is dampened and detergent powder is sprinkled evenly over the area to be cleaned. After a few minutes the floor is scrubbed using a metal fibre or nylon web pad under a scrubbing machine, rinsed well and allowed to dry.

Alkaline powder detergents are particularly effective on terrazzo floors and even badly neglected floors can be effectively cleaned in this way. When used on terrazzo, however, nylon web pads only should be used, as metal fibre pads tend to disintegrate, fibres become trapped in the surface and cause rust to appear.

Alkaline detergents should not be allowed to remain in contact with the skin for any length of time. Hands should always be washed with soap and water at the end of any cleaning operation.

Where skin is found to be particularly sensitive to detergent, rubber or PVC gloves can be used with advantage.

Caustic Materials
Composition

Caustic materials are generally based on caustic soda, in either flake or liquid form, and are extremely strong materials with high pH values. If, to 1 gal of water, 6·5 oz of caustic soda is added, the solution will have a pH value of 14. If 0·65 oz is added, the pH value will be 13 and if only 0·065 oz, or 1·8 g, is added, the pH value of the solution will be 12.

These figures are given to illustrate just how strong caustic soda solutions are.

Uses

Caustic materials are used only where very strong alkaline solutions are required, for example, caustic solutions are used to clear blocked drains. They should never be used on floor coverings because their strong alkalinity can have a damaging effect.

An example of the way which caustic solutions can affect floor coverings occurred in a hospital corridor. The floor, green sheet linoleum, was scrubbed by a new member of the cleaning staff who had not had any training, using a solution of caustic soda in water. After a period of only a few minutes the lino turned a dark yellow-brown colour and no amount of rinsing or further treatment could restore the lino to its original colour.

The only possible course of action that might have met with some success would have been to rinse the floor quickly with an acid solution, in an attempt to reverse the chemical reaction affecting the pigments in the lino. If, however, a delay occurs before the rinsing operation, the chemical reaction affecting the pigments may have progressed so far as to be irreversible.

In another area where a turquoise blue lino turned a pale yellow shade after scrubbing with a very weak solution of caustic soda, the colour was restored almost to the original by rinsing quickly with a solution of 0·1 Normal hydrochloric acid (16 g hydrochloric acid in 1 gal of water). This produced a solution with a pH value of approximately 1 and effectively neutralized the caustic action on the pigment.

It must, however, be stressed that action of this kind, using acid, should only be attempted in an extreme emergency, when drastic treatment is absolutely necessary. Prevention is better than cure and caustic materials should be kept away from all types of floors.

TYPES OF DETERGENT

Acid Cleaners

Composition

Acid cleaners are not used extensively in the cleaning industry, although they are important for some cleaning operations. They do not have as wide a field of application as the neutral and alkaline detergents.

The main materials used in acid cleaners are compounds based on phosphoric acid, sodium bisulphate, oxalic acid, gluconic acid and hydrochloric acid. In most cases surface-active agents are added to assist in the cleaning action.

Acid cleaners are usually aqueous solutions. Formulations are very carefully prepared because of the danger of attack on a wide range of surfaces, including paint, stainless steel, aluminium and almost all floors.

Uses

Acid cleaners must always be used with extreme caution, as they can harm many surfaces. They should always be used under close supervision and their usage and availability always be rigidly controlled.

Phosphoric acid cleaners, in addition to their uses as cleaning and de-rusting agents for steel, are also used in the dairy industry for the removal of a residue known as milkstone.

Acid cleaners are also used for the maintenance of railway coaches and other rolling stock. This is because the dirt on such equipment is high in iron oxide content, which can only be effectively removed with acidic materials. Cleaners used for these purposes are usually aqueous solutions of solid materials, for example sodium bisulphate or oxalic acid, together with surface-active agents and rust inhibitors.

Lavatory bowl cleaners, whether powder or liquid types, are acid in nature. They release acid, often sulphurous acid, when in contact with water and this removes the lime encrustations that build up over a period. Lavatory bowl cleaners of this type should never be used in conjunction with bleach, or hypochlorite, solutions, for when mixed together chlorine gas, which is toxic, is produced.

Weak citric acid solutions have been used for many years for damp-cleaning bronze and, more recently, as a non-streak cleaner for stainless steel.

Dilute hydrochloric acid is widely used for removing lime encrustations from the interior of piping and processing equipment, particularly in hard water areas. It is also used for the removal of cement, plaster and concrete from quarry tiles, an application often necessary in the washroom and toilet areas of new buildings of all types.

If possible, the best way to remove this spillage is to scrape it off by using the edge of a fine carborundum stone. Alternatively, acid can be used. Alkaline materials, such as cement, plaster and concrete, are attacked by acid, and a careful application of a 20% solution of hydrochloric acid in water (i.e. 20 parts acid to 80 parts water) to the affected area might remove the spillage, but if not, a 50% solution in water (i.e. 1 part acid to 1 part water) will remove it.

Hydrochloric acid is very strong and great care must be taken to protect the hands. If any acid should come into contact with the skin the affected area must be washed in soapy water immediately. Acid also attacks shoes and clothing and precautions should be taken accordingly.

The acid solutions should be applied to the affected areas and allowed to remain long enough to soften the spillage, so that it can be scraped off with a scraper or similar tool. If it is allowed to remain on the floor too long it may attack the cement between the tiles. The floor must be thoroughly rinsed several times with clean water afterwards to remove all traces of acid.

If cement, plaster or concrete has been spilled on to terrazzo floors it is better not to use acid unless absolutely necessary, because acid attacks the matrix of the terrazzo and loosens the aggregate. Good results can be obtained using abrasive pads under a polishing machine.

Detergent Crystals

Composition

Detergent crystals, or alkaline degreasers as they are often called, are widely used for industrial applications. They are heavy duty materials, consisting essentially of few ingredients. Perhaps the most important raw material used is sodium metasilicate, which is generally blended with surface-active agents that aid wetting, penetration and removal of loosened dirt.

Sodium metasilicate compositions are soluble in hot or cold water to give a strongly alkaline solution. The solution has pronounced emulsifying and detergent properties and is particularly effective for removing oil, grease and wax.

One of the main advantages of detergent crystals is that they are generally of much lower cost than solvent-based emulsions. Also, because they are solvent-free and based on water only, they can be used on any type of floor.

Uses

Detergent crystals are used mainly for removing accumulations of oil and grease from floors, particularly concrete and asphalt, in garages, factories, driveways, shops and many other locations. They are also used for removing heavy build-up of water emulsion floor wax, in cases where normal alkaline detergents are not sufficiently powerful.

Two methods of use are generally recognized, a hot method and a cold method.

In the hot method, a quantity of detergent crystals are added to water which is as near to boiling as possible. Care should be taken to ensure the crystals are always added to the water and never the water to the crystals. This solution is then applied over the area to be cleaned and allowed to soak into the surface soilage for about half an hour. The floor is then scrubbed with a scrubbing machine, stiff broom or deck scrub. Once the soilage is loosened it should be thoroughly swilled from the floor with clean water and the floor allowed to dry.

The cold method is used when hot water is not readily available.

Loose dirt is brushed off and the detergent crystals sprinkled over the area to be treated. The crystals are then moistened with water, care being taken to ensure the crystals are not rinsed away. The solution should be allowed to penetrate and loosen the dirt and soilage for a minimum of 2 h and preferably overnight. The floor is then scrubbed with a machine, stiff broom or deck scrub, rinsed thoroughly with clean water and allowed to dry.

It is sometimes found that soiled patches re-appear on a concrete surface after it has been cleaned. This is because concrete is porous and is due to the action of the detergent solution bringing oil and grease to the surface. These patches will be removed after two or three applications.

It should be emphasized that care must be taken when using detergent crystals to ensure that colour pigments are fast to the solution. This is particularly important when using them on linoleum, and they must not be allowed to remain on it overnight; the precautions and tests outlined in the section dealing with alkaline detergents should be followed if there is any doubt at all about the stability of the colours in the flooring surface.

Detergent crystal solutions are strongly alkaline in nature and due precautions should be taken to protect the skin and eyes. It is always advisable to wear rubber gloves for skin protection, and should the skin become splashed it should be washed in soap and water immediately.

Solvent-Based Detergent Wax Remover

Composition

Although materials described as solvent-based detergent wax removers are not detergents in the strictest sense, it is convenient to include them under this heading.

They are generally composed of hydrocarbon solvents, often white spirit and water, emulsified together with powerful emulsifying and wetting agents and other additives. The wetting agents have some detergency power and assist in the removal of dirt from the surface being cleaned.

Solvent-based detergent wax removers are manufactured in a wide variety of strengths. In general, however, two types predominate; those consisting almost entirely of solvent with very little water and those with solvent and water in almost equal proportions. Those based almost entirely on solvent are clear, transparent liquids. Those with solvent and water in almost equal proportions are white, opaque liquids resembling milk in appearance. The addition of water to the former produces the milky appearance of the latter.

Uses

Solvent-based detergent wax removers are used mainly for the removal of solvent-based waxes, oil and grease from floors, and oil and grease from equipment and machines.

When used to remove thick, compressed oil or wax from floors, the material should be applied neat and spread with a mop or broom over the affected area. It should then be allowed to soak into the dirt or wax for about 2 h, so that it can penetrate the dirt deposits and loosen them. The area should then be scrubbed using a scrubbing machine in conjunction with coarse grade metal fibre or nylon web pads. If the dirt is particularly thick, for example compacted oil and grease on a concrete floor, scarifying brushes may be required under a machine. If a machine is not available a yard brush or deck scrub dipped in hot water can be used. Exceptionally heavy and stubborn deposits of dirt, oil and grease may require soaking overnight.

After the dirt has been loosened from the floor, it should be removed with a suction drier or mopping equipment and the floor rinsed thoroughly with warm water and allowed to dry.

For routine cleaning of floors contaminated with oil and grease, for example garage floors, a weaker solution should be used. This can be prepared by adding the solvent-based detergent wax remover to water, in proportions recommended by the manufacturer. If the

TYPES OF DETERGENT

dirt deposits are light, the solution need only be allowed to soak into the dirt for a few minutes prior to scrubbing.

These wax removers are widely used for removing paste and liquid types of solvent wax from floors. Constant use of wax can cause build-up, which traps dirt and becomes unsightly. Removal of wax build-up is therefore essential, and periodic wax stripping should be included in every maintenance programme.

When a detergent of this type is applied to a floor, the solvent component penetrates and softens the wax. The emulsifying and wetting agents present hold the wax in suspension and transfer it to the water phase. The loosened wax and dirt can then be easily removed using warm water and a suction drier, or mopping equipment.

When de-waxing a floor prior to sealing, it is particularly important that if mopping equipment is used the water is changed frequently. This is to ensure that all wax is removed and not merely re-applied elsewhere.

It must be recognized that solvent-based detergent wax removers can only be used on floors not harmed by white spirit or similar solvents. They can be used with perfect safety and are recommended for use on wood, wood composition, cork, magnesite, linoleum, concrete and stone floors. They must not, however, be used on asphalt, thermoplastic tiles, PVC(vinyl) asbestos or rubber floors, as these are damaged by solvents.

In the past, paraffin and white spirit, or turps substitute, have often been used for the removal of solvent-based waxes. Solvent-based detergent wax removers, designed specifically for the task of removing oil, grease and wax, have several important advantages over paraffin and white spirit.

First, because of their water content, the solvent-based removers have high flash points and are, in the main, non-flammable.

Secondly, whilst paraffin and white spirit will soften wax, oil and grease and loosen their adhesion to the floor, the absence of any emulsifying agent makes removal very difficult when using a machine or mopping equipment. The result is that the dirt is merely spread around rather than removed.

Thirdly, paraffin and white spirit evaporate quickly from the floor, so that loosened dirt hardens up again. Considerably more material is therefore required to soften the wax, with corresponding increases in cost.

Once a floor has been cleaned using a detergent wax remover, any metal fibre or nylon web pads used should be thoroughly washed in white spirit to remove all dirt, and then allowed to dry.

2

FLOOR SEALS

INTRODUCTION

In recent years great strides forward have been made in floor seal technology. Until comparatively recently the main materials used to seal floors were button polish and, later, oleo-resinous seals. Whilst floor seals represent only about 10% of the total amount of floor maintenance materials produced, a considerable amount of effort has been directed into improving floor seals and methods of their application. The result is that many of those now available are of a very high standard.

Selection of the right type of seal, correct preparation of the floor, application and methods of maintenance are subjects which deserve considerable thought when contemplating sealing a floor. Prior to any floor being sealed, the future of that floor must be taken into account, so that when renewal of worn patches or removal of the seal is required at some future date, the methods of accomplishing them have already been decided.

The number of floors that can be sealed with a conventional solvent-based seal are limited, both with regard to the seal and to the composition of the floor.

The recently introduced water-based seals are playing an increasing part in overcoming the restrictions caused by solvents in seals. Even so, the principle that 'what goes down must come up' should always apply. If a seal cannot be removed from a floor without damaging the floor it should never be applied.

Seals are assuming increasing importance in floor maintenance. When correctly used they have many advantages and therefore knowledge of the different types available, and their characteristics, is essential to anyone responsible for the maintenance of floors.

HISTORY

Seals were first developed for treating porous surfaces, particularly wood and cork. Although timber is, in many respects, an excellent flooring material, liquids can penetrate into many species of timber with comparative ease. Cork, comprised of approximately 50% air, is extremely porous.

HISTORY

It was found that untreated floors quickly absorb dirt and stains, which, once inside the floors are very difficult to remove. Sealing not only prevents dirt and stains from penetrating the floors, but also improves the appearance and makes maintenance easier.

Traditionally, three methods were used to fill the open pores at the surface and thus seal the floors.

First, by oiling. Application of a vegetable or mineral oil was found to increase its water resistance and resistance to stains. The oil also reinforced the surface timber fibres, so that wearing qualities were improved.

The greatest disadvantage of this method is that the oil tends to hold any dust alighting on it. The surface is therefore almost permanently dirty and poor in appearance. A virtue is made of this property by describing the material as 'dust allaying oil'. The oil is sometimes mixed with a hard resin to give a surface with a high degree of resistance to slip. These blends are known as 'gymnasium oil', as they were used mainly in gymnasia.

The requirement for floor oil has now practically disappeared altogether, due to the development of better floor seals. They are, however, occasionally used in areas where absence of dust is of the utmost importance and where the appearance of the floor is of no consequence.

Secondly, waxes were used to 'seal' the floor, and are still used in considerable quantities to date. Waxes are, essentially, soft materials which penetrate into the open spaces and provide a surface which will prevent water and stains from entering the floor. They have the advantage over oil in that they retain their appearance and, when buffed, provide an attractive surface. They are not, however, hard materials and have poor durability compared with conventional floor seals. Even very light traffic can mark waxed floors, or remove the layer of wax. Waxed floors can be easily renovated by maintenance, but this is not always practicable or desirable. Waxes have less resistance to slip than floor seals. They also have poor resistance to stains, and limited resistance to cold and hot water.

In general, domestic premises are best suited to wax. Industrial floors need the more permanent seals, followed by maintenance with wax, rather than wax alone.

Thirdly, as a development from wax, 'button polish', made from shellac resin and methylated spirits was, and still is, used on occasions. Button polish gets its name from the appearance of the pieces of shellac resin, which resemble buttons. This form of polish is generally applied liberally on wood as an initial treatment and a

FLOOR SEALS

thin second coat applied after the first one is dry. The polish seals the open pores of the wood and supplies a surface which will resist penetration by water and stains, and, to some extent, abrasion. Button polish needs protecting with wax, however, as it scratches readily and quickly loses its initial good appearance. This polish is also readily stained by hot water or solvents. It is, however, very quick to dry because of the evaporation of its methylated spirits component.

With the increasing demand for more durable finishes, work on oleo-resinous seals produced a material that was not only better than oil, wax or button polish, but could be easily applied and recoated when required.

In an attempt to increase the durability, and hence lengthen the period before re-sealing became necessary, many other materials were developed. Perhaps the best known are the urea-formaldehydes, both ' one-' and ' two-pot ' and the polyurethanes of various types. From the aspects of durability, speed of drying and chemical resistance, it is generally considered that the polyurethanes are the best types of seal available at the present time.

In very recent times work on water-based floor seals, as distinct from water-based floor waxes, has produced an acrylic seal which is very suitable for use on surfaces which cannot be treated with solvent-based materials, for example thermoplastic tiles, PVC(vinyl) asbestos, rubber and terrazzo. These materials, though not seals in the conventionally accepted sense of being solvent-based, are seals in their own right and play an increasingly important part in modern floor maintenance methods and procedures.

DEFINITIONS

To date, there is no dictionary definition of the word ' seal ', as applied to floor maintenance materials. A floor seal can, however, be defined as ' a semi-permanent material which, when applied to a floor, will prevent the entry of dirt and stains, liquids and foreign matter '.

Seals are not claimed to be ' permanent ' finishes—nothing is permanent. Even buildings are demolished or radically altered in time. If stone floors will eventually crumble and steel stair-treads wear, it is only to be expected that a seal resembling a sheet of plastic material over a floor will also wear.

The earlier exponents imagined that seals were permanent, but time has proved otherwise. Seals are best described as ' semi-permanent ' to indicate as a reminder to the user that at some future date further action will be needed to either remove the seal or repair signs of wear. The word 'semi-permanent' also distinguishes them from other materials that need to be maintained at regular and frequent intervals, for example, the floor waxes. A water emulsion floor wax could also be described as a material which will prevent the entry of dirt and stains, liquids and foreign matter into the floor. Floor waxes, however, need to be maintained at comparatively frequent intervals and cannot, therefore, be included in the category of semi-permanent materials. For this reason they are not considered as 'seals'.

Most seals are based on solvents, although the definition does not specifically state that all seals must be solvent-based. In recent times, water-based seals, as opposed to waxes, have been developed and these comply with the above definition in every respect. Floor seals are, therefore, generally taken to mean solvent-based materials, although one or two types are water-based.

REQUIREMENTS OF A FLOOR SEAL

The requirements of a floor seal are many and varied and depend, to some extent, on the type of floor being sealed. No floor seal yet produced will adequately fill all the requirements, and it is doubtful if such a material will ever become available. However, all floor seals meet certain requirements adequately and some meet most of them quite satisfactorily.

The main requirements of a floor seal are that it should:

    Prevent dirt being ground into the floor surface.
    Protect the surface from water, chemicals and stains due to spillage.
    Provide an attractive appearance.
    Resist scuff marks.
    Possess anti-slip properties.
    Be durable.
    Have good adhesion and resistance to flaking-off.
    Not alter the colour of the floor to which it is applied.
    Not yellow with age.
    Have good levelling and flow characteristics.
    Be easy to patch and repair when required.
    Be easy to recoat.
    Be easily removed, when required.
    Be easy to apply.
    Be quick drying.
    Have a high flash point.

FLOOR SEALS

Have indefinite shelf life.
Have a mild solvent odour.
Have readily available brush cleaning solvent.
Have a realistic cost.

There are, of course, other minor requirements which are not of sufficient importance to warrant consideration here. Those listed above are not in order of importance and priorities will vary according to the circumstances.

A basic requirement of any seal is that it must 'key' to the floor. If, for some reason, the seal does not adhere to the floor it will detach and, under traffic, wear off. However good the quality of a seal, it can only carry out its task successfully if it remains on the floor.

Stiletto heels can cause considerable concern among those responsible for floors; they can quickly ruin soft flooring surfaces, including soft wood and cork. These heels sometimes impose loads of as much as 1200 lb/in^2. Seals applied to such surfaces must not break under the impact and pressure exerted when subjected to stiletto heels. Seals applied in areas which are likely to be trodden by shoes fitted with this type of heel must be flexible, so that the seal will bend and assume the contour of the heel rather than break. No seal will prevent indentations as they are not made of hard, rigid, materials, and are softer than the floors to which they are applied. If the seal bends with the heel the surface remains unbroken and water and dirt cannot penetrate into the floor below. The surface therefore, remains sealed, even though indentations are present (*Figure 2.1*). Conversely, if the seal is not flexible enough

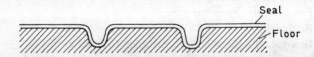

Figure 2.1. Effect of stiletto heels on a flexible seal. The seal bends and assumes the shape of the heel marks without breaking. Despite the indentations, the surface remains sealed

to bend with the heel it will be punctured and any water or dirt can then penetrate through the small hole and result in the floor quickly becoming dirty and very difficult to clean (*Figure 2.2*). In extreme cases, the water may travel along the surface under the seal and eventually lift it from the floor (*Figure 2.3*).

REQUIREMENTS OF A FLOOR SEAL

Perhaps one of the most important requirements of a seal is that it should possess non-slip properties, safety being a very important requirement of any floor, whether sealed or not, especially in hospitals and in homes for aged or infirm people.

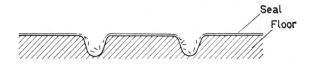

Figure 2.2. Effect of stiletto heels on a brittle seal—Stage I. The seal is punctured by stiletto heels. The surface is no longer sealed and liquids can penetrate into the floor below

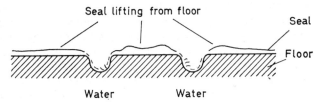

Figure 2.3. Effect of stiletto heels on a brittle seal—Stage II. The seal lifts from the floor due to ingress of liquid into the puncture

Every known and tested floor seal can be termed 'non-slip' in that the coefficient of friction is greater than 0·5 when tested with leather representing the sole of a shoe. The chances of a person slipping when walking normally are therefore less than one in a million.

Gloss is often associated in people's minds with slippery conditions, and because of this, seals giving a full gloss are not always acceptable in certain areas. Often a semi-gloss or eggshell finish is preferred, even though coefficients of friction tests may show that the seals with lower gloss are less non-slip than glossier ones. There is, in fact, no relation whatever between slip and gloss.

Some floors, for example dance floors, sometimes require slippery conditions, although it is understood that highly experienced dancers prefer relatively non-slip conditions. A gymnasium, on the other hand, must have a non-slip finish on the floor. Rather than attempt to control floor friction by adjusting seal formulations, seals are normally designed to be non-slip and if any variation is required, conditions are adjusted with the appropriate floor wax.

TYPES OF SEAL

There is available today a very wide range of floor seals, and new ones are constantly being developed as technology advances. New types of flooring require new materials and methods.

In general, five main types of floor seal are evident:

Oleo-resinous seals
One-pot plastic seals
Two-pot plastic seals
Pigmented seals
Water-based seals

Another material that can be conveniently included under the general heading of 'floor seal' is the silicate dressing, specially formulated for use on large areas of concrete.

Other seals include epoxy esters, styrenated alkyds, modified nitrocellulose, two-pot polyesters, two-pot epoxy resins, styrene/butadiene lacquers, and shellac lacquers. Although some of these seals may possess certain advantages over those included in the main category above (the shellac lacquers, for example, are very quick drying) nearly all have many disadvantages and others have yet to be fully proved in use. They are, however, discussed briefly at the end of this chapter. Each of the types mentioned above is considered in relation to the following characteristics:

Composition
Floors
Shelf life (storage stability)
Colour
Application
Number of coats
Odour
Brush cleaning solvent
Drying time
Intervals between coats
Hardening time
Maintenance
Durability
Chemical resistance
Recoatability
Removability

Oleo-resinous Seals

This type of seal is one of the oldest established and is still probably the most widely used of all. They are relatively cheap, easy to apply and recoat, have no objectionable odour or ill effects on the hands and, when necessary, can be easily renewed.

It should, however, be recognized that oleo-resinous seals are relatively slow drying, rather dark in colour and possess only moderate durability and resistance to chemical attack.

Composition

Oleo-resinous seals consist, essentially, of an oil or oils with resins, solvents and driers. The oils in the earlier types usually included a high proportion of 'tung oil', or 'Chinese wood oil' as it is known. For this reason many of the earlier oleo-resinous seals were often referred to as 'tung oil seals'. Today, however, tung oil is used in smaller quantities, having been replaced by other oils. One blend that gives excellent results is a combination of linseed and dehydrated castor oils.

The resins, which are normally of the phenolic type, and oils are reacted together at high temperature, with solvent, usually white spirit, and driers.

Oleo-resinous seals were often divided into two distinct classes: those with a very low viscosity and those with a high viscosity. Those with a low viscosity were very 'thin' in appearance and were called 'penetrating' seals. They were designed to penetrate and be almost completely absorbed into the floor. Very little remained on the surface, even after two coats. Thus they reinforced the surface layers of the floor to some depth, but the appearance was generally patchy and any gloss soon lost. Low viscosity seals gave good results in that they effectively prevented water and stains from entering the floor in areas where appearance was not of importance.

The high viscosity seals gave a certain degree of penetration into the surface of the floor, although to a lesser extent than the low viscosity seals. A surface layer remained which gave a glossy appearance and a wearing surface. If traffic was light the appearance was generally satisfactory for, perhaps, a year, but if heavy a further coat was required after a few months. This type of seal was generally more acceptable in areas where appearance was important.

Modern oleo-resinous seals are, in general, a compromise between the two. They combine penetration with a surface layer which has an attractive gloss. The durability has lengthened as technology has improved with the result that a high standard can now be obtained.

Floors

Oleo-resinous seals are recommended for use on wood, wood composition, cork and magnesite floors, and should not be used on concrete as the akali in this is liable to break down the seals and cause them to detach from the floor. Neither should they be used on linoleum because of the colour of the seals.

FLOOR SEALS

Oleo-resinous seals are harmful to some floor coverings and must not be used on asphalt, PVC(vinyl) asbestos, flexible PVC, thermoplastic tiles and rubber surfaces, nor are they suitable for use on terrazzo, marble, quarry tiles, stone and other hard surfaces, because of the difficulty of obtaining penetration and hence adhesion, and because of the colour of the seals.

Shelf Life

The shelf life, or storage stability as it is sometimes called, of oleo-resinous seals is very good and they will remain in a satisfactory condition for several years, particularly if the containers are unopened.

A surface skin may form in part-used containers, caused by oxidation of the surface layer of the seal. Any skin must be removed before the seal is used and should never be stirred into the seal as it will only break into small pieces and cause the floor to look unsightly. Old seal which has thickened slightly in the container can be thinned by adding a little white spirit (turps substitute) and stirring well.

Colour

Oleo-resinous seals are generally rather darker in colour than other types of seal since the raw materials are darker; they therefore tend to slightly darken the shade of the floor to which they are applied. This highlights the colours in many of the timber and cork floors, resulting in an improved appearance. Oleo-resinous seals also yellow slightly on ageing, a very gradual process not noticeable on wood and cork floors. For this reason an oleo-resinous seal should never be applied to linoleum.

Application

Mops, lambswool bonnets, roller applicators and brushes are all suitable for applying oleo-resinous seals. Mops should be of the cotton cord type and it is particularly important that one of good quality is used, which will not shed bits into the seal. A lambswool bonnet can be secured to the head of a sponge applicator, or over an old broom stock from which the bristles have been removed.

Roller applicators generally give the best overall results, in terms of speed of application, absence of bubbles and ease of joining two treated areas. Brushes are useful for treating small areas, or areas that cannot be reached with other applicators, for example in corners and under heating pipes.

TYPES OF SEAL

When applying an oleo-resinous seal it is particularly important that a thin even coat be applied, or 'puddles' may form and delay the drying process.

Number of Coats

It is generally recommended that two or three coats are applied, depending upon the type of floor and its condition. While two coats will normally give a satisfactory result on wood, three coats are required to provide a satisfactory surface on cork floors. On the other hand, if a wood floor is particularly absorbent three coats may be necessary.

Odour

The odour of oleo-resinous seals is given mainly by the solvent which is normally white spirit and thus is generally very mild and unobjectionable.

Brush Cleaning Solvent

After using an oleo-resinous seal all applicators and equipment should be thoroughly cleaned in white spirit (turps substitute). When clean, they should be washed in warm, soapy water. This final washing is particularly important for prolonging the life of applicators. The water must not be too hot otherwise brushes can be harmed.

When thoroughly clean, applicators and equipment should be allowed to dry. Brushes should be placed in water, preferably supported so that the weight of the brush is not taken by the bristles.

Drying Time

In general, oleo-resinous seals dry in about 8–10 h. Some take as long as 16 h to dry, but these are the exception rather than the rule. Almost all are hard overnight.

The best possible conditions for drying are warmth and good ventilation, so that evaporated solvent can escape easily.

Intervals between Coats

When two or more coats are being applied a minimum of about 6 h should be allowed between coats. If a second coat is applied too quickly before the first has dried hard, it can lift the first one. There is no maximum interval between coats and additional coats can be applied at any time, providing the surface is clean and dry.

FLOOR SEALS

Hardening Time

Once dry, the seal should be allowed to harden overnight before traffic is allowed over the floor.

Maintenance

As with all types of seal, periodic application of a floor wax will considerably extend the life of an oleo-resinous seal. If the floor is properly sealed, it can be maintained with either a solvent-based wax or a water emulsion floor wax. Water is quite safe to use on a perfectly sealed floor, as it remains on the surface of the seal and never comes into contact with the floor itself. Consideration must, however, be given to the time when the seal will eventually show signs of wear. This may occur in small, localized areas, such as entrances and main traffic lanes. As soon as the seal is broken the floor wax can enter the floor itself; in the case of a wood or cork floor, water may cause loss of colour or splintering. Dirt can also penetrate and cleaning of worn areas may prove extremely difficult.

For these reasons solvent-based waxes are preferred. They are perfectly satisfactory for use on bare wood and cork and are indeed recommended, whereas water emulsion floor waxes are not.

Durability

Durability varies depending upon the condition of the floor, quality of seal, and, of course, type and volume of traffic. In very general terms, however, an oleo-resinous seal should have an effective life under light traffic of about $1\frac{1}{2}$–2 years, and under heavy traffic of about 1 year. The durability in each case should be extended by an additional half to one year if the seal is maintained with a floor wax. These figures are very approximate and are offered only as a guide.

Chemical Resistance

Oleo-resinous seals have only a fair resistance to chemical attack. They are affected by many industrial chemicals, particularly strong acids and alkalis.

They will, however, resist weak solutions of acids and alkalis. Oleo-resinous seals are not recommended for use where chemicals are widely used. Other seals, for example the two-pot polyurethane seals, should be used if there is any danger from chemical attack.

Recoatability

Perhaps one of the best characteristics of an oleo-resinous seal is its recoatability. While many other types of seal, particularly the plastic seals, require careful preparation before they can be recoated, an oleo-resinous seal can be recoated with only the minimum of preparation.

The only requirement is a clean, dry surface, and once the floor is de-waxed, clean and dry, a further coat can be applied, which will adhere satisfactorily to the old seal.

Removability

Oleo-resinous seals are noted for their ease of removal. They can be sanded with a sanding machine or removed with abrasive nylon mesh discs under a polishing/scrubbing machine. Grit 80 discs are recommended. If abrasive nylon mesh discs are used to flat down the surface of the seal, or to remove the top layer only, grit 120 or finer should be used.

Oleo-resinous seals can also be removed by chemical means, using a material similar to a paint stripper. Stripping compounds are based on methylene chloride and because the vapour is rather strong, good ventilation is essential. Stripping an oleo-resinous seal from the floor with chemicals is laborious and should only be attempted if other, mechanical, means cannot be used. Chemical stripping compounds are often used with great success in small inaccessible areas. Rarely are they used on large open areas of floor.

One-pot Plastic Seals

The term ' one-pot ', or ' one-can ', or ' one-pack ', refers to floor seals supplied in a single container. ' One-pot ' is used to distinguish this type of seal from the ' two-pot ' seal, discussed in a later section.

The term ' plastic ', used with reference to a seal, indicates that the seal is composed entirely of synthetic, or man-made materials and does not contain a natural drying oil. Seals of this type dry either by evaporation of solvent or by a chemical reaction which is activated by evaporation of solvent.

Two main types are evident in this class of seal. First, the urea-formaldehyde type, which has in the past made its mark but which is now tending to become obsolescent, and secondly, the polyurethanes, both oil-modified and moisture cured types. While polyurethanes are not ' plastic ' seals in the strictest chemical sense, it is convenient to include them under this heading.

One-pot Urea-formaldehyde Seals

The one-pot urea-formaldehyde seals are generally referred to as ' finishing ' seals, as they tend to remain on the surface and do not penetrate to the same extent as oleo-resinous seals. They dry rather harder than oleo-resinous seals and generally present an attractive appearance in that the gloss is subdued rather than high. They are very light in colour.

Composition

One-pot urea-formaldehyde seals comprise a urea-formaldehyde resin together with an acid catalyst and a blend of solvents. An alkyd resin is often included to improve the flexibility of the seal, as materials without some alkyd present tend to become brittle on ageing. Solvents are rather stronger than those used in oleo-resinous seals as they are required to keep the resin in solution and to hold the acid catalyst in check.

Floors

Wood, wood composition, cork and magnesite floors can be sealed very effectively with one-pot urea-formaldehyde seals.

They should not be used on concrete or linoleum as the finish is rather too hard. They are harmful to asphalt, PVC(vinyl) asbestos, flexible PVC, thermoplastic tiles and rubber surfaces, because of the action of the solvents. They are not recommended for treating terrazzo, marble, quarry tile, stone and other similar, hard surfaces, owing to the difficulties in obtaining satisfactory adhesion. These seals also have a tendency to become brittle on ageing and may be scratched on very hard floors.

Shelf Life

The presence of acid catalyst in the same container as the reactive resin component can cause stability problems in storage. The acid is held in check by a strong, very volatile solvent. On application, the solvent evaporates, allows the acid to react with the resin and the film hardens and dries.

The formulation must be very delicately balanced. On the one hand, if too little acid is present the seal will not dry hard when it is applied; on the other hand, if too much acid is present a reaction may take place in the container, resulting in the seal thickening and, in extreme cases, becoming solid. The shelf life can, therefore, vary depending upon the formulation, from as little as a few months to as long as, perhaps, 2 years.

Shelf life can be affected by the state of the container. If there is a leak at the lid, solvent may escape over a long period, so allowing the hardening reaction to start. Containers should, therefore, always be well closed if it is intended to store the seal for any length of time.

Containers that are not full need special care, because there will be a considerable amount of air space above the contents relative to the surface area. Thus, solvent can evaporate, allowing the material to solidify in the container.

It is always preferable to decant any seal from a part-full container into another, smaller, container, which can then be closed tightly with as little free air space as possible. In this way the seal will have the longest possible shelf life.

Colour

One-pot urea-formaldehyde seals are noted for their very light colour, which they retain extremely well on ageing, and do not have the tendency to yellow, as is often the case with some other types of seal. They are therefore very suitable for use on even the lightest coloured floors.

Application

Lambswool bonnets, roller applicators and brushes are the most suitable applicators, and of these, rollers applicators are preferred because of their speed in applying the seal. As with other seals, the material should be spread thinly and evenly over the floor. If a thick film is applied, the drying time may be considerably lengthened.

Number of Coats

Two or three coats are generally recommended, depending upon the porosity of the floor and the type and volume of traffic over it. Porous floors may appear patchy after two coats and a third coat may be essential to ensure an even finish. Two coats should, however, produce a satisfactory result on hard floors.

Odour

Solvent odour is rather stronger than that of oleo-resinous seals, because of the stronger solvents used. Adequate ventilation is therefore essential to enable evaporating solvents to escape and to reduce their odour to a minimum.

Brush Cleaning Solvent

Brushes or applicators should be cleaned immediately after use with thinners recommended by the manufacturer of the seal used.

FLOOR SEALS

White spirit, or turpentine substitute, is not sufficiently strong and should not be used.

Drying Time

One-pot urea-formaldehyde seals dry, in general, quicker than oleo-resinous seals and are usually hard dry in 6–8 h.

The rate of drying is largely controlled by the rapidity with which the solvent can evaporate from the floor, so allowing the acid component to react with the resin. Warm, well ventilated conditions are ideal. If ventilation is poor and solvent cannot escape, the acid will be held in check and will be unable to react, so that drying will be retarded.

Intervals between Coats

A minimum of about 6 h should be allowed between coats. As with the oleo-resinous seals, a second coat applied too soon may lift the first coat and ruin the whole sealing operation.

Whilst there is no maximum interval between coats, if a seal is allowed to harden for about 2 weeks or more before a further coat is applied, the floor should be roughened to remove all gloss. It should also be lightly mopped with approved solvent immediately prior to application of the new seal. This will soften the old seal slightly, thus providing the best possible 'key' for the new seal.

Hardening Time

The final coat of seal should always be allowed to harden overnight before traffic is allowed over the floor.

Maintenance

One-pot urea-formaldehyde seals should be maintained with a floor wax, either solvent-based or water emulsion. As with the oleo-resinous seals, solvent-based waxes are preferred.

Durability

It is extremely difficult to give an indication of the durability that can be expected from a one-pot urea-formaldehyde seal, since, not only do traffic conditions differ from place to place, but formulations vary widely also.

In general, however, durability similar to that given by oleo-resinous seals can be expected. Nevertheless, it must be stressed that whilst some one-pot urea-formaldehyde seals may not achieve the life of some oleo-resinous seals, others may show a greatly improved durability, particularly if well maintained.

TYPES OF SEAL

Chemical resistance

The chemical resistance of one-pot urea-formaldehyde seals is very similar to that of the oleo-resinous seals.

The seals will withstand normal cleaning agents and some of the weaker chemicals, but not the stronger chemicals, particularly strong acids and alkalis. They should not, therefore, be used if chemical resistance is one of the main requirements of the seal being applied.

Recoatability

Renewal of worn patches is rather more difficult than with oleo-resinous seals, as dried films of urea-formaldehyde tend to resist solvents more than do oleo-resinous films.

An area requiring a further coat must be roughened to provide a mechanical key and damp mopped with the appropriate solvent to slightly soften the film and provide a chemical key, before a fresh coat is applied. The new application should be made to the centre of the worn area and feathered out to the edges, so that it blends with the remainder of the seal.

Removability

One-pot urea-formaldehyde seals are more difficult to remove than oleo-resinous seals because they dry to a harder finish. They can be sanded quite satisfactorily, or flatted down and removed, using abrasive nylon mesh discs under an electric polishing/scrubbing machine. Grit 80 is recommended.

Chemical stripping compounds, based on methylene chloride, will also remove these seals. Two applications may be required. Because of the rather strong odour given by the vapour, good ventilation is essential.

One-pot Polyurethane Seals

Two different types of seal are included under this heading. First, the oil-modified polyurethane, also known as a urethane-oil, and secondly, the moisture-cured polyurethane.

The oil-modified type is not, in the strictest chemical sense, a member of the one-pot plastic seal group, as some oil is present. It is, however, convenient to include it in this category, together with the moisture-cured polyurethane, which is a true polyurethane and also a plastic seal.

Both types of polyurethane give improved performances in almost every respect over the oleo-resinous and one-pot urea-formaldehyde seals. The key to their success is their ability to form hard, yet flexible, coatings.

FLOOR SEALS

The one-pot polyurethane seals are rapidly increasing in usage, in many cases at the expense of the one-pot urea-formaldehyde seals. They are easy to apply, give good, consistent results and combine improved chemical resistance with extended durability.

Composition

The oil-modified type consists of a pre-reacted polyurethane together with a small proportion of oil, solvents and additives. The drying mechanism is, therefore, very similar to that of the oleo-resinous seals.

The moisture-cured type, on the other hand, consists of pure polyurethane together with solvents and additives. The isocyanate constituent absorbs moisture from the atmosphere and reacts with it, hardening to form the polyurethane film. In all but extremely arid climates there is sufficient moisture in the atmosphere to dry the film within a few hours. Small variations in humidity are not sufficient to retard the drying mechanism of the seal.

Floors

Both types of polyurethane seal are recommended for use on wood, wood composition, cork and magnesite floors; on concrete, the moisture-cured polyurethane will give better results. The oil-modified type is not generally recommended for use on concrete as the alkali present in the concrete can cause loss of adhesion after a short period.

Neither type is recommended for use on linoleum, because of the difficulties of later removal. Nor are they recommended for use on asphalt, PVC(vinyl) asbestos, flexible PVC, thermoplastic tiles and rubber surfaces, because of the possible damaging effects of the solvents, or on terrazzo, marble, quarry tiles, stone and other hard surfaces because of the difficulty of obtaining satisfactory penetration, resulting in lack of adhesion.

Shelf Life

The shelf life of the oil-modified polyurethane seal is almost indefinite, as the polyurethane itself is already fully reacted. However, there has been, in the past, some difficulty in ensuring a reasonable shelf life (a period of several months) for the moisture-cured polyurethane, since it is very sensitive to any moisture present in the container, or in the atmosphere. A seal of this type will react with any moisture entering at the lid, or at a pin-hole leak anywhere in the container, causing the seal to thicken and, in extreme cases, to solidify.

This defect has now largely been overcome by the addition of special additives to the seal, which absorb any moisture present. Modern moisture-cured polyurethane seals can remain in a stable, liquid state, for an extended period of time, sometimes for 1 or 2 years.

Moisture-cured polyurethanes should never be allowed to remain in part-full containers, as they will absorb moisture and thicken. Any remaining should be decanted into a smaller container so that the ullage, or free air space, is minimized. The container should then be tightly lidded and the tin momentarily inverted, to seal any small pin hole leak that may be present.

Colour

Both types of polyurethane seal are light in colour, the moisture-cured being the lighter of the two. The colour of the oil-modified polyurethane lies about halfway between the oleo-resinous and the one-pot urea-formaldehyde seals. The moisture-cured polyurethane resembles the urea-formaldehyde seal in colour and is very suitable for application to light-coloured floors.

Application

Both types can be applied by brush, lambswool bonnet and roller applicator, the last-mentioned probably being the best method.

Whichever applicator is used, the seal must be applied thinly and evenly. This is particularly important in the case of the moisture-cured polyurethane, to ensure that the whole depth of the film is dry before any further seal is applied.

Number of coats

Two or three coats of either type should be applied, depending upon the nature and porosity of the floor. On cork particularly, three coats will probably be required to provide a smooth surface with an even finish.

Odour

The solvents used in both types of seal are generally of the white spirit or xylene type. Odours are, therefore, generally mild and inoffensive.

Brush-cleaning solvent

Brushes and applicators must be cleaned immediately after use, particularly when a moisture-cured polyurethane seal has been used, because the seal will absorb moisture from the atmosphere and harden if left for a few hours. In the case of oil-modified polyurethane, brushes and applicators can be cleaned quite effectively with white spirit. A special, stronger solvent is necessary to clean tools used for the application of a moisture-cured polyurethane and a solvent recommended by the manufacturers of the seal should be used.

Drying Time

Oil-modified polyurethane seals dry hard in about 3–4 h. The drying mechanism (evaporation of solvent and oxidation of the oil constituent) is very similar to that of the oleo-resinous seals. Warmth and good ventilation tend to accelerate the drying process.

Moisture-cured polyurethane seals take a little longer to dry, the normal drying time being about 6–8 h or even more. Seals of this type can be formulated to dry quicker, but this usually results in a greatly reduced shelf life.

Intervals between Coats

In the case of the oil-modified polyurethane seal, a minimum of about 3–4 h should be allowed between coats. The first coat must be dry before a further one is applied.

The maximum interval between coats is in the region of 24 h. If a longer period is allowed to elapse, the earlier application may harden to such an extent that penetration and therefore adhesion of the new coat to the old, is impaired.

A coat of moisture-cured polyurethane seal should be allowed to harden for a minimum of 8 h before a further coat is applied. Any further coats should be applied within about 2 days otherwise inter-coat adhesion may be lost.

Hardening Time

After the final coat has been applied, both types of polyurethane should be allowed to harden overnight before traffic is allowed on the floor.

Maintenance

The useful life of both types of seal will be considerably extended if the floor is maintained with a wax. Whilst water emulsion waxes can be used with confidence on a well-sealed wood floor, the solvent type is preferred, because when the seal eventually wears the newly applied wax will come into contact with the floor itself, and whereas solvent wax is satisfactory on bare wood a water emulsion floor wax may cause discolouration or splintering.

Durability

In general, the expected durability of an oil-modified polyurethane seal lies somewhere between that of an oleo-resinous seal and a two-pot polyurethane seal. Each formulation varies and durability will depend on the quality of the seal, the type and volume of traffic, and on subsequent methods of maintenance.

As an estimate, an oil-modified polyurethane seal should have an effective life of about $1\frac{1}{2}$–$2\frac{1}{2}$ years under light traffic and about 6 months less under heavy traffic. If the seal is maintained with a floor wax the durability should be extended by about a year. It should be noted that many instances are known of oil-modified seals remaining in a perfectly satisfactory condition even after 4 or 5 years, when properly maintained.

Moisture-cured polyurethanes are generally more lasting than the oil-modified seals, because they are manufactured from pure polyurethane materials and do not include any oil, which is less durable. In general, however, they do not have as long a life as the best two-pot polyurethane seals, because, owing to difficulties in can stability, the solid content of the moisture-cured material is normally lower.

Chemical Resistance

The chemical resistance of oil-modified polyurethanes is better than that of the oleo-resinous and urea-formaldehyde seals, but not as good as the two-pot polyurethane seals. They will resist dilute solutions of acids and alkalis in addition to petrol, oil, salt water and grease and are quite suitable for use on floors where mild chemical attack can be expected. They will not, however, resist attack by concentrated solutions of acids and alkalis.

The moisture-cured polyurethanes have better chemical resistant properties because of the absence of oil. They will resist a wide variety of chemicals and common reagents and are attacked only by the more concentrated solutions of chemicals, particularly acids and alkalis.

Recoatability

Recoatability of both types of seal is more difficult than with oleo-resinous seals and more care must be taken to ensure a successful result.

Oil-modified and moisture-cured polyurethanes harden over a period of time and become more chemically resistant with age. They will therefore resist a fresh application of seal, and inter-coat adhesion may not be obtained if the floor is not correctly prepared.

Correct preparation in both cases consists of roughening the old seal to remove all gloss and provide a mechanical key for the new seal. This can be carried out using abrasive nylon mesh discs, grit 120, under an electric polishing/scrubbing machine. The floor should then be thoroughly cleaned and mopped with solvent recommended by the manufacturers before new seal is applied. The purpose of mopping with solvent is to slightly soften the old seal sufficiently to enable the new one to key to it. The new seal should be applied as soon as the solvent has evaporated.

By preparing the floor in this way the new application has the best possible chance of adhering mechanically and chemically to the old seal.

Removability

As the moisture-cured polyurethane dries harder than the oil-modified, it will, after a period, be more difficult to remove.

Removal of both types should, preferably, be by sanding. If only partial removal is required, this can be carried out using abrasive nylon mesh discs under an electric polishing/scrubbing machine. Grit 80 or 100 is recommended, finishing with grit 120 to provide an even surface for the new application of seal.

Whilst chemical stripping compounds, based on methylene chloride, may remove an oil-modified polyurethane seal, they have little effect on moisture-cured seals and are not generally recommended on seals of this type.

Two-pot Plastic Seals

The description 'two-pot plastic seal' is given to those seals which require the blending together of two component parts prior to use. The components are a 'base' which comprises the main body of the seal, and an 'accelerator' or 'hardener' which, when added to the base, initiates a chemical reaction that only stops when the material has fully hardened.

The principle of hardening a resinous film by means of an accelerator represents an advance over the one-pot materials which dry by solvent evaporation or film oxidation. Better quality materials can be manufactured by the two-pot method without any loss of storage stability.

It should be emphasized, however, that bases and accelerators must be mixed in the correct proportions, in accordance with the manufacturers' instructions. If they are not mixed in the right proportions, the seal will either dry slowly and become soft, if accelerator is omitted, or dry too quickly with the resultant loss of penetration and adhesion, if too much accelerator is added. For the convenience of the user, almost all, if not all manufacturers supply two-pot materials in the correct mixing proportions. The base is usually supplied in a larger container than the accelerator. Sufficient space is allowed in the base container to take all the accelerator, so that when the latter is added to the base the total volume just fills the container. The lid should then be replaced. and the container shaken to ensure complete and thorough mixing. Alternatively, if lever lid, open top containers are used, the material can be thoroughly stirred, using a stick or similar article. Stirring must be rigorous to ensure the accelerator is evenly dispersed throughout the base. The importance of thorough mixing cannot be over-emphasized, as poor mixing will lead to patchy drying and a poor finish.

As soon as the accelerator is added to the base the hardening reaction starts. It is therefore essential that the seal is blended just prior to use and is not allowed to stand for some time in a mixed condition. If the seal is mixed too soon it will thicken and may be found to be unusable when required.

The period during which the material is usable once the base and accelerator have been blended together is termed the ' pot-life ' of the seal. The pot life will vary according to the formulation of the seal from as short as 3–4 h to as long as 2–3 days.

Ideally, therefore, when using a two-pot seal, all preparation should be completed before the seal is mixed. Mixing should be carried out immediately before use and the seal allowed to stand for about 5 min to enable any air bubbles that have been introduced into the seal during mixing to float to the surface and escape. If these air bubbles remain in the seal when it is applied, they may become trapped as the surface dries, resulting in a rough, unsightly appearance, resembling sandpaper. In this event, the only remedy is to lightly sand the seal with a sanding machine, or remove the surface layer with abrasive nylon mesh discs under an electric polishing/scrubbing machine, and apply a further coat of seal.

Whilst correct and thorough surface preparation is an essential requirement for success for any sealing operation, it is particularly relevant where two-pot seals are concerned. As the seals dry quickly (in some cases they are hard dry within 2 h), they have not the time to penetrate to the same extent as the slower-drying seals. It is therefore imperative that the floor is thoroughly clean and dry to ensure maximum adhesion and hence durability.

Results achieved with a two-pot seal amply reward the extra care required. Durability and appearance are excellent and chemical resistance is also very good. The fast drying times that can be obtained with the better quality products reduce inconvenience to a minimum.

Extra care and supervision are, however, required, particularly when mixing accelerator with base. Several instances are known where the wrong accelerator has been added, giving poor results. On one occasion the accelerator was omitted altogether, later to be returned to the manufacturer for credit!

Two-pot plastic seals generally have a strong odour because of the types of solvent required in their manufacture. Good ventilation is therefore essential. Subsequent removal may present some difficulty due to the excellent adhesion obtained, but if the sealed floor is maintained correctly, ' touching-up ' of worn areas in heavy traffic lanes should not be necessary for a number of years.

Two main types of two-pot plastic seal are available today. First, the two-pot urea-formaldehyde seals, which give a considerably improved performance over the one-pot urea-formaldehyde seals. Secondly, the two-pot polyurethane seals, which, in addition to having excellent hardness and abrasion resistance, are particularly recommended for their exceptional durability and resistance to chemical attack. They are also noted for their low flammability.

Two-pot Urea-formaldehyde Seals

Two-pot urea-formaldehyde seals are basically ' finishing ' seals. They dry to a higher gloss than is obtained with the one-pot urea-formaldehyde type and have an attractive appearance but are now largely superseded by the two-pot polyurethane seals, which are better in almost every respect.

Composition

Their composition is very similar to that of the one-pot urea-formaldehyde seals, except that the reactive components are supplied in separate containers.

The base consists of urea-formaldehyde resins in solvents, usually with an alkyd included to improve the flexibility of the seal. The accelerator consists of an acid, often in industrial methylated spirits or another, similar, alcohol. When the accelerator is added to the base a reaction starts which continues until the material has fully hardened.

Floors

Two-pot urea-formaldehyde seals are recommended for use on wood, wood composition, cork and magnesite floors, but are not

TYPES OF SEAL

recommended for use on concrete, as better materials are available, nor on linoleum, as they tend to become too hard and brittle for this type of floor covering. They are harmful to, and must not be used on asphalt, PVC(vinyl) asbestos, flexible PVC, thermoplastic tiles and rubber surfaces. They are unsuitable for use on hard surfaces, such as marble, terrazzo, quarry tiles and stone, because of the difficulties in obtaining penetration and hence adhesion.

In the case of terrazzo floors, instances have occurred where floors have been cleaned with metal fibre floor pads and fragments have remained ingrained in the terrazzo. Acid accelerator, spilt onto the floor, rusted the metal causing unsightly brown rust spots to appear under the seal. Spots so formed are very difficult to remove; the only really satisfactory way being to remove the seal, using abrasive nylon mesh discs under an electric polishing/scrubbing machine, and extract the fragments and rust spots using a special poultice formulated for the purpose.

Shelf Life

Since it is very stable and cannot react on its own, the shelf life of the base component is normally very good and the material can be stored indefinitely.

The shelf life of the accelerator is also very good, provided the pack is satisfactory. As the accelerator is basically an acid, the pack must be able to resist acid attack. Glass and polythene containers are generally quite satisfactory for accelerators.

Colour

Two-pot urea-formaldehyde seals are very light in colour, being almost water-white. They are, perhaps, the lightest coloured of all the wide range of solvent-based seals, and because of this, are very suitable for use on light coloured wood floors, having the further advantage that they will not yellow over a long period of time.

Application

Roller applicators, lambswool bonnets and brushes are all suitable for applying two-pot urea-formaldehyde seals.

It is particularly important to ensure that the coats are spread thinly and evenly over the floor because two-pot seals of this type have a higher solids content than the one-pot type, and should a thick coat be applied, some difficulties in penetration into the surface and slow drying may result.

FLOOR SEALS

Number of Coats

In general, two coats will give a satisfactory result. On very porous floors however, it may be necessary to apply three coats to obtain an even finish.

Odour

The solvents used in two-pot urea-formaldehyde seals are, of necessity, strong, therefore a strong smell of solvent will occur whilst the material is drying, and ventilation must be adequate to reduce the odour to a minimum. Once the seal is dry and all evaporating solvent has been removed, no odour remains.

Brush Cleaning Solvent

As with the one-pot urea-formaldehyde seals, special solvent is required to clean equipment because white spirit and other similar materials of this type are not strong enough. The solvent recommended by the manufacturers of the seal should always be used.

Drying Time

Two-pot urea-formaldehyde seals generally dry hard in about 3–4 h. Warm, well-ventilated conditions will assist drying considerably.

Intervals between Coats

The minimum interval between coats is about 6 h which enables the first coat to dry really hard before the next is applied. If a second coat is applied too soon, solvent could penetrate the undried first coat and lift it from the surface.

Whilst there is no maximum interval between coats, a seal which has been allowed to harden for about 2 weeks or more should be roughened to remove all traces of gloss and mopped with the appropriate solvent immediately prior to the application of seal.

Hardening Time

After sealing, the floor should be allowed to harden overnight before traffic is allowed over the floor.

Maintenance

Once sealed, the floor should be maintained with a floor wax to protect the seal from wear. A water emulsion floor wax can be satisfactorily used for this purpose, but a solvent-based floor wax is preferable.

TYPES OF SEAL

Durability

The durability of two-pot urea-formaldehyde seals is generally better than that of oleo-resinous and one-pot urea-formaldehyde seals, but not as good as the two-pot polyurethane seals.

As with all seals, the durability depends on the type and volume of traffic together with subsequent maintenance. Periodic application of a floor wax will considerably extend its life, in some cases for a matter of years.

Chemical Resistance

Whilst the two-pot urea-formaldehyde seal exhibits better chemical resistance than the one-pot type, it has not as good a resistance to chemicals as the two-pot polyurethane seal, showing some degree of blistering, discolouration and loss of adhesion when attacked by grease, whereas the latter remain unaffected. Strong solutions of alkaline materials will also discolour the two-pot urea-formaldehyde type of seal, while the two-pot polyurethane seal resists strong alkaline materials well.

Recoatability

Touching-up worn patches is generally rather more difficult with two-pot seals than with one-pot seals because the former usually have a greater film thickness and are harder. This also applies to two-pot urea-formaldehyde seals. It is, therefore, always preferable to re-treat areas showing signs of wear before the seal is worn right through to the floor. Once a seal has worn completely through, more drastic treatment, such as a partial sanding operation, may be necessary.

An area requiring retreatment must be roughened to remove all gloss, thus providing a mechanical key for the new seal. It should then be damp mopped with an appropriate solvent and seal applied as soon as the solvent has evaporated. This will provide the best possible conditions to enable the new seal to adhere to the old. The new seal should be applied to the centre of the area being treated and feathered out to the edges.

Removability

Two-pot urea-formaldehyde seals should be removed by sanding, using a sanding machine. Partial removal, if required, can be carried out by using abrasive nylon mesh discs under an electric polishing/scrubbing machine. Grit 80 discs are recommended for this purpose.

Chemical stripping compounds will remove two-pot urea-formaldehyde seals with only limited success. Two or more applications may be required to remove the old seal.

Two-pot Polyurethane Clear Seals

Two pot polyurethane seals have been developed comparatively recently and have now largely superseded the two-pot urea-formaldehyde seals. With regard to durability and chemical resistance, they are probably the best seals available at the present time.

The best types are pure polyurethanes, unmodified with additional materials which tend to weaken the film. Quick-drying types can be walked on in a few hours and three coats can be satisfactorily applied in less than 1 day.

Composition

A two-pot polyurethane seal consists of a base, which is a type of polyester, and an isocyanate accelerator. Both materials are held in a stable solution in strong solvents. When the isocyanate is added to the polyester a chemical reaction takes place resulting in a tough, but extremely flexible, polyurethane film.

It should be noted that the accelerator is extremely sensitive to moisture in the atmosphere and the container should be kept well sealed at all times. If it comes into contact with the skin it must be washed off immediately with soap and water.

Floors

Two-pot polyurethane seals are recommended for use on wood, wood composition, cork and magnesite floors on which surfaces they give excellent results, particularly with regard to appearance, durability and chemical resistance.

They can also be used satisfactorily on concrete floors, but the appearance is rather patchy, particularly when using a clear seal. Better, more attractive results can be obtained from a pigmented, i.e. coloured, two-pot polyurethane seal.

Provided certain precautions are taken, two-pot polyurethane seals can be made to give excellent results on asphalt floors, which are normally damaged by solvents. Details regarding the precautions that must be taken are given in Chapter 3.

Two-pot polyurethane seals must not be used on thermoplastic tiles, PVC(vinyl) asbestos, flexible PVC and rubber floors, nor on hard surfaces such as marble, terrazzo, quarry tiles and stone, because of the difficulties in obtaining satisfactory adhesion to these floors.

TYPES OF SEAL

For some time there has been controversy regarding the sealing of linoleum with a two-pot polyurethane seal. Whilst there are many instances of floors having been sealed successfully with a seal of this type, in general, the use of a solvent-based seal on linoleum is not recommended because of the difficulties of touching-up worn patches and later removal. On balance, it is considered that although sealing linoleum with a two-pot polyurethane seal can prove advantageous, regular waxing with a water emulsion floor wax or solvent-based wax is to be preferred.

Shelf Life

The base component has an indefinite shelf life that should extend over a period of several years. The accelerator, however, consisting essentially of a highly reactive isocyanate, is extremely sensitive to water and moisture vapour. It will, if allowed, extract moisture from the atmosphere and form a hard crust on the surface. In extreme cases, for example if the lid has been left off the container, the accelerator will quickly harden throughout and become solid. This solid, or fully reacted material, is useless and must not be used. If accelerator is found to be even partly reacted it must not be added to the base, but should be exchanged for fresh, liquid material.

The shelf life of the accelerator will therefore depend largely on the state of the container. If even a pinhole leak is present in the container some thickening may occur after a very short time.

Accelerator should always be stored in metal, usually tin-plate containers; if stored in glass it tends to thicken however well the glass is sealed. This is because ingredients in the glass trigger off the hardening reaction.

Colour

Two-pot polyurethane seals are very light in colour, a trait they retain well over a period of years, and are suitable for even the lightest coloured floors.

Application

After mixing the base and accelerator, the seal can be applied by roller applicator, lambswool bonnet or brush. A roller applicator is preferred to a lambswool bonnet and this method should always be used to seal large areas of floor, because two-pot polyurethane seals have a tendency to trap air and form bubbles, particularly during application. Lambswool bonnets, due to their construction, tend to hold small pockets of air. When the bonnets are dipped in seal the air becomes mixed with the seal and bubbles are formed.

Also, having completed a stroke, the bonnet must be lifted from the floor. This action tends to cause the formation of a ridge of bubbles. Although these bubbles should burst and the seal flow out evenly, if the solvent evaporates quickly from the surface the bubbles may be trapped under a layer of seal resulting in a rough, unsightly appearance.

A roller applicator, on the other hand, is used in conjunction with a tray and all air can be displaced from the roller before application begins. At the end of a stroke, the roller can be raised gently, whilst still turning, and thus prevent a ridge of bubbles being formed. As with all seals, a thin, even coat should be applied.

Number of Coats

A minimum of three coats is generally recommended to give the best results. The first, or priming coat, is usually supplied with a lower solids content and lower viscosity than the finishing coats. This is to enable the first coat to penetrate the surface and provide a key for subsequent, or finishing coats which provide the hard-wearing surface. If a priming coat is not used penetration, and therefore adhesion and durability, may be impaired.

In heavily trafficked areas, for example entrances and main corridors, a further coat of finishing seal will prove advantageous. This should be applied after the priming coat and before the finishing coats are applied overall.

Odour

The solvents used in the manufacture of two-pot polyurethane seals are rather strong and the odour can be somewhat pungent. It is, therefore, recommended that areas being sealed should be well ventilated to allow solvents to escape and to assist drying.

Once dry, there is of course, no residual odour.

Brush Cleaning Solvent

Special solvent is necessary to clean brushes and applicators after use, and the solvent recommended by the seal manufacturer should be used.

It is very important that all equipment used is thoroughly cleaned immediately after use because the chemical hardening reaction is continuing and the longer the equipment is left the harder it will be to clean. On no account should it be allowed to soak in water, as the latter will simply hasten the hardening process. Particular attention should be given to the stocks of brushes.

TYPES OF SEAL

After cleaning in the approved solvent, all items should be washed in a solution of a neutral detergent or soap and water.

Drying Time

The early types of two-pot polyurethane seal dried hard in about 6–8 h; however, with modern types the drying time is very much shortened and the quick-drying seals are hard dry, suitable for overcoating, in approximately 2 h.

Seals can be made to dry even faster, but it has been found that problems can arise through this practice due to the shortened pot life and resultant thickening in the container.

A two-hour drying time appears to give the best possible balance between speed of drying, pot life and ease of application.

Intervals between Coats

The minimum interval between coats is about 2 h, assuming that a seal is being used with a two-hour drying time.

The maximum interval is in the region of 24 h and any further coats should be applied within this period, because two-pot polyurethane seals rapidly harden and develop a resistance to chemicals.

Ideally, successive coats should follow each other at intervals of 2 h, so that all coats in a system are applied in the shortest possible time. Satisfactory results are obtained if, in a three coat system, two coats are applied one day and the third coat the next. If, however, a longer interval is allowed before the final coat is applied, the surface should be roughened and mopped over with the appropriate solvent before seal is applied.

Hardening Time

After the last coat has been applied, a minimum of 4 h hardening time should be allowed before traffic is permitted on the floor. If possible, the seal should be allowed to harden overnight.

Maintenance

Once dry, the seal will continue to harden for a period of about 14 days, after which it will attain its maximum abrasion and chemical resistance. For about 4 or 5 days after sealing it is advisable not to allow water to come into contact with the floor. Maintenance should be carried out dry, or with a solvent-based liquid wax.

After the first 4 or 5 days the surface can be maintained with either a water emulsion floor wax or, preferably, a solvent-based wax.

FLOOR SEALS

Durability

Two-pot polyurethane seals are, in all probability, the most durable seals available. The durability is excellent by any standards and when properly maintained, seals can remain in an unworn condition for many years. Thus the period between recoating is considerably extended, resulting in a minimum of disruption and inconvenience. The floor also remains in more attractive and satisfactory condition over a longer period.

As a rough guide, under light traffic conditions a two-pot polyurethane seal should remain in good condition for a period of about 3–4 years. If maintained with a floor wax, a further 1–2 years or more can safely be added.

If the traffic is heavy, the seal should remain in good condition for a period of about 2 years, or 2–3 years or more if maintained with a floor wax.

Chemical Resistance

Two-pot polyurethane seals exhibit perhaps the best chemical resistance of all seals, resisting attack by solutions of acids and alkalis, grease, petrol, oil and other common chemicals. They are, therefore, eminently suitable for use in areas where chemical attack can be expected, for example, the floors of laboratories, chemical factories, hospital wards, garages and food factories.

Recoatability

Because two-pot polyurethane seals are extremely hard and possess exceptional adhesion properties, they are difficult to recoat, and to remove. It is always recommended that areas showing signs of wear should be recoated before the seal is allowed to wear through to the floor itself.

The area to be recoated should be roughened, using abrasive nylon mesh discs, grit 100 or 120, under an electric polishing/scrubbing machine. The floor should be mopped with the appropriate solvent and the seal applied as soon as the solvent has evaporated. For best results the seal should be applied to the centre of the area and feathered out to the edges.

Removability

Sanding is the only really effective method of removing a two-pot polyurethane seal. It can be partially removed, if necessary, using abrasive nylon mesh discs, under an electric polishing/scrubbing machine. Grit 80 discs are recommended for this purpose.

Chemical strippers of the methylene chloride type are not strong enough to be effective.

Pigmented Seals

Pigmented seals are those that contain a pigment, or colour. They resemble paint in appearance and are not to be confused with clear seals to which a stain has been added. They are used to impart a pleasing or uniform colour to floors and to strengthen the surface; they are widely used on concrete and magnesite floors, and also, to a lesser extent, on asphalt. Some of the main areas of application are in corridors, stores, boiler houses, factory working areas, ambulance bays, warehouses and garages.

These seals differ from conventional paints in that they are specially formulated for use on floors. Most paints are primarily designed for protective and decorative purposes, for outdoor use. They are, therefore, formulated on oils, alkyd and other varnishes that will withstand the weather and will expand and contract readily, without cracking, as the temperature varies. They are not designed specifically to resist abrasion, nor to resist attack by alkaline materials that are present in concrete. If a conventional paint is applied to concrete it will rapidly show signs of wear and will also be affected by the alkali. The varnish constituent in the paint is, literally, converted into a soap, or is 'saponified', and the paint will break down and flake off.

Pigmented seals, on the other hand, are formulated to resist abrasion by foot and other traffic and to resist attack by alkaline materials found in concrete. They are, therefore, very much more durable and are more suitable for use on concrete than conventional paint in every respect.

Two main types of pigmented floor seal are evident. The first is based on synthetic rubber resins and is a one-pot (ready for use) material. This gives a pleasing appearance, has good durability and resists many forms of chemical attack. The second is based on polyurethane materials and it possesses physical and chemical properties very similar to those of the clear polyurethane seals. This type of seal is superior in all respects to those based on synthetic rubber resins, except for simplicity, as the best pigmented polyurethane seals are two-pot materials.

Synthetic Rubber Pigmented Seals

Pigmented seals based on synthetic rubber have been readily available for a number of years. One of their main advantages, in addition to their simplicity, is the ease with which further coats can be applied, as required.

They are generally cheaper than the two-pot polyurethane type and give a good all-round performance.

Composition

This type of seal consists of coloured pigments in a medium of rubber resins and solvent.

The pigments are chosen for, among other qualities, their complete resistance to attack by alkaline materials. The seal is, therefore, perfectly satisfactory for application to old concrete. New concrete must be allowed to harden for at least 14 days and there must be no evidence of water in the pores. Floors treated with a synthetic rubber pigmented seal will remain completely unaffected by any alkali present in the concrete over a period of many years.

Floors

Synthetic rubber pigmented seals are recommended for use on concrete and magnesite floors, and can also be used satisfactorily on some stone floors, providing there is sufficient porosity to enable the seal to key satisfactorily to the floors. A modified type, using slightly faster evaporating solvents, can be used and is recommended for use on asphalt. Because asphalt varies widely in its composition a trial area should always be treated first to ensure that the desired results are obtained.

Seals of this type must not be applied to thermoplastic tiles, PVC(vinyl) asbestos, flexible PVC, rubber or linoleum surfaces. It is also inadvisable to apply them to hard, impervious surfaces, such as terrazzo, marble and quarry tiles.

Shelf Life

The shelf life, or storage stability, is generally very good and the seal can be stored satisfactorily over a period of several years. It is inevitable that during any period of storage some settlement of pigment will occur because its specific gravity is greater than that of the medium. Therefore, before using a synthetic rubber pigmented seal it is essential that it is thoroughly stirred to ensure that all pigment is evenly distributed and that none remains on the bottom of the container. Soft settlement can be easily dispersed by stirring, ideally, with a large palette knife or, alternatively, a piece of wood. If, however, settlement is heavy, it will be easier to disperse if some of the liquid is first decanted into another container. The settlement should then be stirred vigorously and the decanted liquid returned slowly, with stirring.

Colour

These seals are generally available in a range of colours. Perhaps the most popular being a dark shade of red, sometimes called 'Tile Red', and 'Grey'. Both colours are very serviceable and are pleasing to the eye.

Application

Brushes, lambswool bonnets and roller applicators are satisfactory for applying a synthetic rubber pigmented seal, which should be applied thinly and evenly over the floor. If the floor is rough and uneven, for example, rough concrete, application by lambswool bonnet or roller applicator alone may be difficult because the applicator may not be able to reach indentations and hollows in the floor, thus leaving the seal only on the high spots.

On a rough floor of this type, seal should be spread over an area about 18 in wide, using an applicator, then spread evenly and forced into the hollows with either a 4 in brush or a turk's head brush. The latter, on a handle, is more convenient if a large area is to be treated. This method ensures that the whole floor is effectively sealed and that small iregularities in it are not missed.

Number of Coats

It is recommended that at least two coats should be applied. One is generally insufficient, particularly on a porous floor, and two coats will give a good durable finish.

A third coat can always be applied with advantage on the main traffic lanes and in the entrances. The main areas subjected to heavy traffic should be treated first, followed by two coats over the entire floor area.

Odour

Solvents used in the manufacture of synthetic rubber pigmented seals are not pungent in any way; consequently, whilst drying, their odour is mild and unobjectionable.

Brush Cleaning Solvent

Special solvent is required for cleaning applicators and equipment, and that recommended by the manufacturers of the seal should be used.

FLOOR SEALS

After thoroughly cleaning all applicators and equipment with the approved solvent, they should be washed in warm, soapy water and allowed to dry.

Drying Time

The drying time ranges from about 3–8 h, depending on the formulation of the seal. Other factors which greatly influence the drying time are thickness of application, condition of the floor, temperature and state of ventilation. A thin, even film on an old, hard floor in warm, well-ventilated conditions are the best possible circumstances for rapid drying.

Intervals between Coats

A minimum of about 6 h should be allowed between coats. If a second is applied too quickly it may soften the first one and result in slow drying.

There is no maximum interval between coats. One of the main advantages of synthetic rubber seals is that they can be over-coated at any time.

Hardening Time

A sealed floor should be allowed to harden overnight before traffic is allowed on to it.

Maintenance

The floor should be swept in the normal manner. Day-to-day soilage can be removed by damp mopping, using warm water. Heavy traffic areas should be cleaned periodically with a solution of a neutral detergent in water.

It is rarely necessary to wax pigmented seals, due to the nature and use of the floor itself. If, however, such a course is decided on, a water emulsion floor wax should be used. Solvent-based waxes must be avoided, as solvents soften rubber-based seals.

Durability

Durability of synthetic rubber pigmented seals is good, although not as good as the two-pot polyurethane type.

Under light traffic, two coats would be expected to remain in good condition for a period of about $1\frac{1}{2}$–2 years. Under heavy traffic, this would probably be reduced to about 1 year.

These figures are given solely as a guide as durability depends on many widely differing factors. Numerous instances are known of

TYPES OF SEAL

seals giving at least twice the durability of the figures above. The estimates are, however, based on the averages of many case histories.

Chemical Resistance

Synthetic rubber pigmented seals have, in general, good resistance to chemical attack. They are, in particular, very resistant to oil and grease and can be used with absolute confidence on areas of floor likely to be affected by them. They are also resistant to hot water, dilute acids and alkalis. Films are, however, softened by solvents of the white spirit or petroleum types.

Recoatability

Ease of touching up and ready recoatability are two important characteristics of synthetic rubber pigmented seals. Each new application of seal softens the old one, with the result that the new and old applications blend and key together.

No special preparation is necessary, other than ensuring that the floor is clean and dry. Any oil or grease must be removed using detergent crystals, or a solvent-based detergent wax remover, before new seal is applied.

Removability

If, at any time, it is desired to remove a synthetic rubber pigmented seal from magnesite, this can be effectively carried out using a sanding machine. Alternatively, abrasive nylon mesh discs, grit 80, can be used under an electric polishing/scrubbing machine. The latter method can also be used for removal of seal from asphalt.

Removal from smooth concrete can be effected by sanding, or by using abrasive nylon mesh discs or a scarifying brush under a polishing/scrubbing machine. If, however, the concrete is ribbed or uneven, it can only be removed by solvent. This is a difficult, laborious task and should only be attempted in small areas at a time.

White spirit, or a stronger solvent, is spread onto the area to be treated and allowed to soak into the seal for several minutes. Loosened seal is then removed with mopping equipment, using a bucket of solvent. The solvent must be changed frequently to ensure that seal is removed from the floor and not merely re-applied elsewhere. Several applications of solvent may be necessary to remove the seal.

Two-pot Polyurethane Pigmented Seal

Two-pot polyurethane pigmented seals have, like the clear seals, been developed comparatively recently. They are, in all probability, the best available in the range of pigmented seals, particularly with regard to durability and resistance to chemical attack. Their properties are very similar to those of the two-pot polyurethane clear seals. The formulations of the pigmented seals have generally been modified to overcome problems associated with concrete, magnesite and asphalt floors.

Composition

A two-pot polyurethane pigmented seal consists of a base component and an accelerator component. The base comprises a type of polyester and includes the pigment, or coloured, constituent and special additives to resist saponification by the alkali present in concrete.

The accelerator comprises an isocyanate in solvent and is generally clear, that is, not pigmented. It is very sensitive to moisture and, if allowed, will extract moisture from the atmosphere, thicken and become useless. It must, therefore, be stored in a well sealed container at all times.

When the accelerator is added to the base a chemical reaction takes place, resulting in a tough, yet flexible, polyurethane film. If any accelerator is allowed to come into contact with the skin it must be washed off immediately with soap and water.

Floors

The seals are particularly suitable for use on concrete and magnesite floors. Because of the alkali resisting materials included in the best formulations, they are admirably suited to withstand attack by alkali present in concrete.

Some types of two-pot polyurethane pigmented seal can be used very satisfactorily on asphalt floors, provided certain precautions are taken, detailed in Chapter 3. Many such floors have been sealed using this type of material and have withstood heavy traffic over a period of years.

The seals can be used on some types of stone, providing it is porous enough to allow penetration and, therefore, adhesion. It is always advisable to carry out a test in cases of doubt.

The seals must not be used on thermoplastic tiles, PVC(vinyl) asbestos, flexible PVC, linoleum and rubber floors, and are generally unsuitable for use on hard surfaces such as marble, terrazzo and quarry tiles, because of the difficulties in obtaining adhesion to them.

TYPES OF SEAL

Shelf Life

The shelf life of the base component is indefinite and the material should remain in good condition over a period of years. As the pigment is included in the base it is likely that some settlement will occur. This must be thoroughly dispersed in the medium when the seal is used. If settlement is allowed to remain on the bottom of the container, opacity, or hiding power, of the seal will be considerably reduced.

The accelerator, however, being extremely sensitive to moisture, must be kept in a tightly closed container at all times. Remarks concerning the shelf life of two-pot polyurethane clear seal accelerators, detailed above, also apply in this case.

Colour

Two-pot polyurethane pigmented seals can be manufactured in almost any colour. They are, however, restricted to a few of the more popular shades.

Perhaps the most popular, as with the synthetic rubber pigmented seals, are a dark red, or ' Tile Red ' and ' Grey '. Black is sometimes required, mainly for use on asphalt floors and yellow and white are often used for marking lines.

Application

Application can be carried out using a brush, lambswool applicator or, on concrete, a turk's head brush. The seal is, in general, too low in viscosity to be applied satisfactorily using a roller applicator as this tends to cause splashing.

On an uneven surface, for example a ridged and pitted concrete floor, a lambswool bonnet can be used in conjunction with a brush, as described in the section dealing with the application of synthetic rubber pigmented seals, above.

Number of Coats

A minimum of three coats is generally recommended to give the best results. The first, or priming coat, is usually supplied at a lower solids content and lower viscosity than the finishing coats. This is to enable the first to penetrate the surface and provide a key for the subsequent coats which provide the hard, wearing surface and also the colour.

Odour

As with all true polyurethane seals, the solvents necessary to keep the resins in solution are relatively strong. Odour, therefore,

FLOOR SEALS

can be pungent, although not as powerful as that associated with clear two-pot polyurethane seals, as the total volume of solvent is less. Areas being sealed should be well ventilated to allow solvent to escape and to assist drying. Once dry, no residual odour remains.

Brush Cleaning Solvent

Special solvent is necessary to clean brushes and equipment and that recommended by the manufacturers should always be used.

All equipment must be thoroughly cleaned immediately after use because the chemical hardening reaction continues and the longer the equipment is left the harder the cleaning task will be. Water hastens this hardening process, and should not be allowed to come into contact with equipment until this has been cleaned with solvent.

Particular attention should be given to the stocks of brushes. After cleaning equipment with solvent, all items should be washed in a solution of a neutral detergent, or soap, in water.

Drying Time

The drying time is generally very similar to that of the clear two-pot polyurethane seals, described above. The better quality, quick drying seals are hard dry in about 2 h.

Intervals between Coats

The minimum is approximately 2 h, assuming that a seal is being used with a two-hour drying time. The maximum interval is about 24 h and any further coats should be applied within this period.

As with the two-pot polyurethane clear seals, successive coats should, ideally, follow each other at two-hourly intervals. Satisfactory results are, however, obtained if, in a three coat system. two coats are applied one day and the third the next. When a longer interval is allowed before the final coat is applied, the surface should be roughened and mopped over with the appropriate solvent before seal is applied.

Hardening Time

Once all coats have been applied, a minimum of 4 h hardening time should be allowed before traffic is permitted on the floor. If possible, the seal should be allowed to harden overnight.

Maintenance

The floor should be maintained by brushing and periodic damp mopping, using a solution of a neutral detergent in water.

TYPES OF SEAL

Pigmented seals are rarely waxed, but the use of a floor wax will considerably extend the life of the seal. Either a solvent-based or a water emulsion floor wax can be used for this purpose.

Durability

The durability of two-pot polyurethane pigmented seals is excellent. Under light traffic, a life of about 2–3 years, or longer, can be expected. If traffic is heavy, the floor can be expected to remain in good condition for at least $1\frac{1}{2}$–2 years. Many instances are known of floors remaining in an excellent condition after much longer periods than those indicated. Durability does, however, depend upon a number of factors and the figures quoted are only offered as a very rough guide.

Chemical Resistance

The chemical resistance properties of two-pot polyurethane pigmented seals are excellent. They will withstand attack by solutions of acids and alkalis, grease, petrol, oil, water and many other common chemicals.

They are, therefore, suitable for use in areas liable to be subjected to chemical attack. Whereas synthetic rubber pigmented seals are softened by petrol, the polyurethane seals are not and thus they can be used in garages and on areas around petrol pumps in addition to warehouses, laboratory floors, chemical and food factories, corridors and many other places.

Recoatability

Two-pot polyurethane pigmented seals are extremely hard and develop exceptional adhesion properties. They are difficult to recoat and very difficult to remove. As with clear polyurethane seals, areas showing signs of wear should be recoated before the seal is allowed to wear through to the floor itself.

An area to be recoated should be roughened to remove all gloss, using abrasive nylon mesh discs under an electric polishing/scrubbing machine. The floor should be mopped with an approved solvent before a new coat of seal is applied. As soon as the solvent has evaporated, the seal should be applied to the centre of the area and feathered out to the edges.

Removability

Sanding is the only really effective method of removing a two-pot polyurethane pigmented seal. Alternatively, the use of grit 80 abrasive nylon mesh discs may achieve the required results if the floor is level.

Chemical strippers are not very effective on this type of seal.

Water-based Seals

Water-based floor seals are now widely used although they have been in existence for only a few years.

Their usefulness is greatest in the restoration of worn floors that cannot be effectively treated with a conventional, solvent-based seal. Examples of floors on which water-based seals have proved of great value are thermoplastic tiles, PVC(vinyl) asbestos, porous linoleum, rubber, terrazzo and asphalt.

Prior to the introduction of water-based seals, surfaces such as worn thermoplastic tiles, rubber and asphalt were treated with a floor dressing consisting essentially of resins in alcohol solvents. Such solvents will not dissolve these floors. A dressing of this type could, therefore, be used with perfect safety as far as damage to the floors is concerned.

They have, however, a number of disadvantages. They dry so quickly that they are difficult to apply without leaving brush or applicator marks. They are highly flammable and their use increases fire risk. They form brittle films which tend to crack and chip on ageing and they must be protected by some form of wax dressing. Some resins used in alcohol floor dressings are emulsified by alkaline materials and are at least partially removed at each wax stripping operation.

Recent developments in water-based seal technology have overcome the objections found when using alcohol-based floor dressings. They flow evenly and well, are non-flammable, do not become brittle on ageing and are not removed by alkaline materials during routine cleaning operations.

It should be recognized that the seals lack the durability, coat for coat, of conventional solvent-based seals. Because of this they must be maintained with a water emulsion floor wax. In the right circumstances, however, water-based seals will not only improve greatly the appearance of a floor, but will also reduce maintenance costs very considerably. They have the added advantage that they can be easily touched-up, or removed, as required.

The advent of the water-based seal has proved to be a major advance in floor maintenance technology.

TYPES OF SEAL

Composition

Water-based seals are supplied as ready-for-use materials. The best quality seals consist of acrylic polymer resins together with a small amount of plasticizer to assist flexibility. The acrylic polymer resins are of large particle size, in contrast to those used in the manufacture of water emulsion floor waxes, which are of small particle size.

The large particle size polymer resins fill the open pores in the floor being treated, so providing a protective 'plastic skin' over the surface.

They are formulated to be resistant to neutral and alkaline detergents and are not affected by them. When periodic stripping of the water emulsion wax used to maintain the floor is necessary, the seal remains unaffected.

Floors

Water-based seals are designed for use on thermoplastic tiles, PVC(vinyl) asbestos, flexible PVC, rubber, porous linoleum, terrazzo, marble and asphalt floors.

An example of their usefulness has been demonstrated on old battleship grade linoleum, maintained over many years with detergent and water only. The linoleum was very porous and four coats of a conventional water emulsion floor wax were required to produce a satisfactory surface. After each stripping, using an alkaline detergent solution, four new coats were required, an expensive operation in both labour and materials.

After sealing with two coats of a water-based seal only two coats of floor wax were required. When the floor was later stripped, using an alkaline detergent, the seal remained intact on the floor and only two new coats of wax were needed to give a satisfactory finish.

The application of a water-based seal to terrazzo and marble floors fills the open pores and provides a smooth surface, so preventing dirt from entering the floor, and making maintenance easier.

Asphalt, always a difficult surface to maintain in first class condition, has perhaps benefitted most of all from the advent of a water-based seal. Old asphalt surfaces tend to become dull in appearance and coloured asphalt often fades. Very large areas of asphalt have been revived by application of a suitably coloured water-based seal. This also provides a sound base for further maintenance with a coloured water emulsion floor wax, which enhances **the effect.**

The seals can be used on magnesite and some stone floors. They can also be used on concrete, although better, more durable, seals are available.

Water-based seals must not be used on wood, wood composition or cork floors, not only because of the water in the seals but also because they tend to break down and assume an unsightly, white, powdery appearance.

Shelf Life

The shelf life of water-based seals is very similar to that of conventional water emulsion floor waxes. The seals should remain in their original condition for approximately 1 year. After this period some thickening may take place due to chemical changes in the seal.

The seals should never be exposed to extremes of temperature and must be protected from frost.

Colour

Water-based seals are water-white in colour, because of the very large proportion of acrylic polymer resin present.

Once applied, they retain their very light colour over a long period of time and do not yellow on ageing. This is of great importance when white, or very light coloured, tiles are being treated. Lightness in colour is one of the outstanding features of these seals.

Application

The seals can be applied using a lambswool applicator or mopping equipment, exactly as conventional water emulsion floor waxes.

The coverage of about 3000–4000 ft^2/gal is again similar to that obtained with conventional water emulsion floor waxes. It is very much greater than that which can be obtained with solvent-based floor seals. A thin, even coat should be applied.

Number of Coats

The number of coats will depend largely on the porosity of the floor. One or two are generally sufficient to seal the surface satisfactorily for further treatment with a water emulsion floor wax.

Odour

The odour given by water-based seals is much milder than those associated with solvent-based seals. This is because the solvent is, in this case, water, which is odourless. The odour given by a drying seal is very similar to that of an acrylic emulsion floor wax, widely used today.

TYPES OF SEAL

Cleaning Solvent

After use, all equipment should be cleaned in warm, soapy water, thoroughly rinsed and allowed to dry.

It is important to clean all equipment immediately after use, otherwise the resins will set hard and be very difficult to remove.

Drying Time

Water-based seals dry hard in approximately 20–30 min, depending on the temperature, humidity and thinness of film applied. A second coat can be applied as soon as the first is dry.

Intervals between Coats

The minimum interval between coats is approximately $\frac{1}{2}$ h. The first coat must be dry before a further coat is applied.

There is no maximum interval between coats and a further coat can be applied at any time. It is, of course, essential that before this is done the surface must be clean and free from all traces of wax and grease.

Hardening Time

Traffic can be allowed over the floor as soon as the seal is dry, usually after about 30 min. It is, however, desirable to allow the seal to thoroughly harden for at least an hour, or longer if possible, so that it will be in a better condition to take traffic.

Maintenance

Water-based seals are not designed to resist constant abrasion by foot traffic, but rather to fill open pores in a floor and provide a basis for waxing. They must, therefore, be protected with a floor wax. As the seal is water-based the wax must be of the water emulsion type and not solvent-based.

The floor should be waxed as soon as the seal has dried hard. Periodic stripping of the wax with an alkaline detergent will not harm the seal which is formulated to resist such materials.

Durability

The durability of water-based seals depends on the type and volume of traffic, but perhaps even more on the method of maintenance.

If the floor wax is allowed to wear off the seal will be exposed to abrasive foot traffic and will deteriorate quickly. If, however, the floor wax is well maintained, the seal itself should never be subjected to abrasion. Under these conditions it will last very much longer, and can be expected to remain in good condition for a period of about 1–3 years. After this period, when the seal shows signs of wear, a further coat will be necessary.

Chemical Resistance

Acrylic water-based seals have some chemical resistance. They will resist water, oil, grease and neutral and alkaline detergents, but will not withstand attack by acids or solvents.

Recoatability

Water-based seals are very easy to recoat and a new application will adhere well to the old seal.

The floor must be completely clean and free from all traces of dirt, oil and grease. All water emulsion floor wax used to maintain the floor must also be removed. For this reason, any touching-up or recoating should be carried out during the periodic wax stripping operation.

When touching-up worn areas only, seal should be applied to the middle of the area and feathered out to the edges. The new application will key to the old and provide a satisfactory surface for further maintenance with a water emulsion floor wax.

Removability

If, at any time, it is desired to remove a water-based seal, this can be carried out using an alkaline detergent solution incorporating a solvent. A material of this type is sufficiently strong to break down the seal film but not strong enough to harm the floor. The solution should be applied to the floor and after being allowed to act for about 5 min, scrubbed using coarse grade metal fibre or nylon web pads under an electric polishing/scrubbing machine.

Alternatively, the seal can be removed by mechanical abrasion only. Abrasive nylon mesh discs, grit 120 or finer, should be used under an electric polishing/scrubbing machine. A coarser grit than 120 should not be used as discs may leave swirl marks on the floor. If grit 120 discs, or finer, are used, the surface will remain perfectly smooth. Any floor can be satisfactorily treated in this way, including rubber.

The floor should be re-sealed immediately all the old seal has been removed.

Silicate Dressings

Silicate dressings are widely used on large areas of concrete, where it is not a practical proposition to apply a conventional solvent-based floor seal because of the expense. Application of a silicate dressing has the advantage of being a relatively cheap method of treating large areas. The dressing reinforces the surface of the concrete and prevents it from dusting.

Composition

Silicate dressings consist, essentially, of sodium silicate dissolved in water with additives mixed into the blend. The dressing functions in two ways. First, the sodium silicate reacts with lime in the concrete to form insoluble calcium silicate. Secondly, it loses water by evaporation and forms silicate glass.

The result is that the surface is chemically hardened, dust free and resistant to a wide range of chemicals.

Floors

Silicate dressings are designed for use on concrete floors, although they can be used with advantage on some stone floors.

They cannot be used on wood, wood composition or cork floors, nor on floor coverings, and are also unsuitable for use on terrazzo, marble and quarry tiles.

Shelf Life

The shelf life of silicate dressings is almost indefinite and the material will remain in a satisfactory condition for a period of years.

Even though the material is water-based it can be stored in metal containers without rusting taking place because the silicate itself acts as a metal preservative and prevents the containers from rusting.

Colour

The dressings are supplied in different colours depending upon the manufacturer. It is normal to colour the material slightly so that it can be readily recognized. Once on the floor the material assumes a colourless appearance.

Application

Silicate dressings are applied by making a solution of the material in the quantity of water recommended by the manufacturer. The solution is then sprayed onto the floor through the rose of a watering can, until saturation point is reached.

Any unabsorbed solution should be brushed on to an adjoining, untreated area, using a soft hair sweeping brush. If any unabsorbed material is allowed to dry on the surface it will form a glassy layer.

FLOOR SEALS

Silicate dressings can be applied to new concrete as soon as it has hardened sufficiently to walk on. If possible, the concrete should be allowed to harden for about 7 to 14 days before being treated. Old concrete can be treated at any time, providing the surface is clean and free from dirt, particularly oil.

Number of Coats

Concrete floors vary greatly in porosity and the number of coats required to give a satisfactory result will vary from floor to floor. In general, two or three coats of diluted material in water will be required.

Odour

Silicate dressings are odourless because the only material evaporating from the floor is water. They can therefore be applied even in work areas whilst people are carrying out their normal duties.

Brush Cleaning Solvent

All equipment should be cleaned after use in warm water containing a little neutral detergent, or soap. After cleaning, equipment should be allowed to dry.

Drying Time

Each coat will be surface dry in about 1 h and hard dry in about 12–14 h.

Intervals between Coats

The minimum interval between coats is about 12–14 h, to allow each coat to fully react with the concrete floor. There is no maximum interval and further coats can be applied as and when required.

Hardening Time

After the last coat has been applied, the floor should harden overnight before traffic is allowed on the floor.

Maintenance

The floor should be maintained by sweeping. Day-to-day soilage can be removed by damp mopping, using a solution of a neutral detergent in water.

Durability

The durability of a silicate dressing is very difficult to assess, because it depends not only on the type and volume of traffic, but also on the state of repair of the concrete floor and on the treatment given. Under normal conditions, wear in the main traffic lanes generally becomes evident after periods varying from 6 to 18 months. If the concrete is in good condition when silicate dressing is applied, the floor may remain in an excellent condition for a period of several years, particularly in areas not subjected to heavy traffic.

Chemical Resistance

Silicate dressings have limited chemical resistance properties. They will, however, withstand the effects of water, oil, petrol and similar materials.

They are affected by acids and strong alkaline solutions and these should be prevented from coming into contact with the dressing.

Recoatability

Worn areas are extremely easy to recoat. Once the floor is clean and free from all traces of dirt, oil and foreign matter, application can be made using a watering can and soft hair sweeping brush, as described above. The new coat will satisfactorily key to the old surface.

Removability

Silicate dressings are very difficult to remove from concrete floors as they are chemically reacted with the surface of the concrete and form part of it. For this reason, removal using conventional methods and materials should not be attempted or the concrete may be damaged in the process. If, however, removal of the dressing is necessary, specialist advice should be obtained.

Other Seals

Included under this heading are the following types of seal:

Epoxy esters.
Styrenated alkyds.
Modified nitrocellulose.
Two-pot polyesters.
Two-pot epoxy resins.
Styrene/butadiene lacquers.
Shellac lacquers.

The above list is by no means complete and many other materials, sometimes wrongly termed ' seals ' are available. They are, perhaps, the more widely known materials and each will be discussed briefly.

Epoxy Esters

Epoxy ester seals consist, essentially, of the same raw materials that are used in oleo-resinous seals, except that an epoxy ester resin is used in place of the phenolic resin normally present in oleo-resinous seals.

The result of the substitution is that the seals dry quicker and are generally harder. Gloss and other physical properties are very similar to those of conventional oleo-resinous seals.

Epoxy ester seals are not manufactured on a very wide scale. This is probably due to the disproportionate increase in raw material costs compared with performance.

Styrenated Alkyds

Styrenated alkyd seals are to be compared with oleo-resinous seals and have certain advantages over them. They are quicker drying, slightly lighter in colour and the durability is probably marginally improved.

They have, however, a disadvantage in that they are difficult to recoat. Strong solvents are necessary in some styrenated alkyds to keep the resin in solution. When recoated, the strong solvents penetrate the old seal and tend to lift it from the surface.

Modified Nitrocellulose

The main advantage of seals based on nitrocellulose is the rapid speed of drying that can be achieved with these materials. Strong, volatile solvents are necessary to keep the nitrocellulose in solution. On application, the solvents evaporate rapidly and as drying is effected solely by solvent evaporation the seals will dry in about $\frac{1}{2}$ h, or even less under favourable drying conditions.

The nitrocellulose film has not, however, the chemical resistance or durability of many of the seals mentioned above. The solvent odour is also pungent. Nitrocellulose seals have a tendency to become brittle on ageing and show scratch marks rather easily.

Two-pot Polyesters

Two-pot polyesters are widely used in the furniture trade as high quality finishes. They set very rapidly initially, then continue hardening for an extended period of time.

They are only used to a small extent as floor seals. This is because they eventually harden to such an extent that they have a tendency to become brittle and scratch easily.

The raw materials from which two-pot polyesters are made are expensive and the seals themselves are, therefore, priced rather

TYPES OF SEAL

higher than is generally expected of floor seals. As they do not, in general, perform as well as the two-pot polyurethane floor seals, which are in the same approximate price range, the latter are preferred.

Two-pot Epoxy Resins

Two-pot epoxy resin seals have been compared, on a cost/performance basis, with two-pot polyurethane seals.

They give extremely hard and very durable finishes. Perhaps their most outstanding characteristic is their excellent resistance to chemicals of all types. Because of this, they are widely used in the paint and surface coating industry for treating such widely differing items as silos, vats, chemical containers and storage tanks in the holds of ships. Excellent adhesion to metal surfaces is another important characteristic.

They have the disadvantage with regard to their use as floor seals, however, that they dry rather more slowly than two-pot polyurethane seals. The drying time can be considerably extended at low temperatures and may be doubled or trebled in cold, wintery conditions. Thus, it has been found in practice that two-pot polyurethane materials are generally preferred for sealing floors. The two-pot epoxy materials have their main use in the paint industry for the manufacture of chemically resistant surface coatings.

Styrene/Butadiene Lacquers

Whilst styrene/butadiene lacquers have been used for many years with considerable success in the paint industry, they are rarely included in the formulations for floor seals.

Perhaps the main disadvantage is that the material does not dry sufficiently hard to resist abrasion by foot traffic. It has, therefore, a tendency to show marks and to wear rather quicker than most other types of seal.

Shellac Lacquers

Shellac lacquers generally consist of shellac blended with other resins and plasticizers in an alcohol solvent medium. They have the advantage of drying very rapidly, by evaporation of solvent.

One of the earliest shellac lacquers is known as ' button polish '. Button polish (see page 27) scratches readily and quickly loses its initial good appearance. It is, however, still used on occasions.

Later developments were formulated to give film thicknesses much greater than those obtained with the conventional button polish but have, however, the disadvantages usually associated with this polish. They rapidly become brittle, scratch easily and are readily marked by water and alcohol solvents.

Shellac lacquers are not, therefore, to be generally recommended, particularly as they are comparatively costly to maintain.

FACTORS AFFECTING THE CHOICE OF SEAL

Sealing a floor can be a costly operation but one sealed with the correct material can enable the initial cost to be recovered many times over by reduced maintenance costs. In addition, the floor presents a more attractive appearance.

A floor could, however, be ruined by application of the wrong type of seal; several instances are known of thermoplastic and PVC(vinyl) asbestos tiles being treated with solvent-based seals, resulting in softened tiles and colours running into each other. In these cases the floor tiles had to be lifted and the floor relaid.

Before any seal is purchased it is strongly recommended that a number of factors relating to sealing the floor are considered. A wide variety of seals are available and the choice of the best one for a given set of conditions is often difficult. Too often a seal is chosen hastily, sometimes with little, if any, thought. It may become obvious, after a short time, that it was not the best for the job. Once applied, it is very difficult to change the seal without completely removing it from the floor by sanding or other means.

A material which has given every satisfaction on one floor may not be the best for use on another, even though both floors may be similar. This is because circumstances may vary, particularly with regard to traffic conditions. Each type of seal has its own peculiar characteristics, advantages and disadvantages and these must be fully considered before any is specified.

The properties of the main types of seal have already been discussed. Consideration will now be given to the most important factors affecting the choice of seal. These are listed below, not in any order of priority, except that the first one, concerning the aim of sealing the floor, must be resolved first. Each of the other factors should then be considered, if only briefly. When all relevant factors have been considered, the important ones, i.e. those that will influence most the choice of seal, can be determined.

Factors affecting the choice of seal

Reason for sealing—the aim.
Type of floor (e.g., wood, terrazzo).
Appearance required.
Present method of maintenance.

FACTORS AFFECTING THE CHOICE OF SEAL

Seal, if any, at present on the floor.
Availability of floor for resealing at a later date.
Durability of seal.
Drying time.
Solvent odour.
Chemical resistance.
Resistance to yellowing on ageing.
Flammability.
Slip resistance.
Flexibility.
Recoatability.
Removability.
Subsequent method of maintenance.
Advisability of using a priming coat.
Cost, initial and over an extended period.

Each of the above factors will now be considered in turn.

Reason for Sealing—the Aim

The most important factor that must be considered is the reason for sealing the floor. If, for example, it is to be sealed to prevent deterioration due to spillage of chemicals, then only a chemically resistant floor seal, such as a two-pot polyurethane, should be used. If the aim is to brighten up a concrete floor, a pigmented seal should be used.

A critical examination of the aim may reveal that an alternative method of maintenance, perhaps waxing, would be preferable. Not all floors should be sealed and in many cases a thorough waxing is to be preferred.

The reason for sealing the floor must be clear before any further factors are considered.

Type of Floor

The type of floor under consideration is another very important factor. Some floors, for example thermoplastic tiles, PVC(vinyl) asbestos and rubber, must not be sealed with conventional solvent-based seals because the solvents present in the seals have a detrimental effect upon these surfaces and soften the floors. Only water-based seals can, therefore, be used on these floors.

Similarly, with other floor surfaces only certain types of seal can be used.

Wood, wood composition and cork are examples of floors which can be sealed with a variety of materials. Oleo-resinous, one-pot plastic or two-pot plastic seals can be used on these floors and consideration would need to be given to other factors before a choice could be made.

Appearance Required

When considering appearance it must be recognized that penetrating seals of the oleo-resinous type do not give as high a gloss as the two-pot plastic seals and the latter should be used if a high gloss is required.

Concrete floors can be greatly improved in appearance by the use of a pigmented seal. The choice lies, therefore, between a synthetic rubber and a two-pot polyurethane pigmented seal. If appearance is not as important as the prevention of dust, a silicate dressing may give the required results, at a much lower cost.

In general, the better the appearance required, the better the quality of the seal that must be used.

Present Method of Maintenance

Where maintenance has been effectively carried out it is likely that the minimum of preparation will be required prior to sealing. If maintenance has been neglected, however, a considerable amount of prior work may be necessary. If no maintenance has been carried out over a long period of time, it may not be advisable to seal the floor at all.

The floors of some old mills, for example, consist of wood block. Oil from machines has been allowed to drip onto the blocks for many years, with the result that they are saturated throughout with oil. No amount of sanding would remove the oil and it is extremely doubtful whether any seal would adhere satisfactorily to a floor in this condition.

Seal, if Any, at Present on the Floor

If the floor has not been sealed, a wider choice of material is generally available than if it has. Not all seals are compatible and some cannot be applied on top of others.

For example, if a floor has been sealed with an oleo-resinous seal it can only be resealed with a similar material. If a different type, say a two-pot polyurethane is applied, the new seal will lift the old. Again, if a floor has been treated with a two-pot plastic seal and an

FACTORS AFFECTING THE CHOICE OF SEAL

oleo-resinous seal is applied on top of it, the latter will not adhere to the two-pot plastic seal and will soon chip and flake off.

Resealing of floors is considered in greater detail below.

Availability of Floor for Sealing at a Later Date

Some floors, for example, those in schools, can be made readily available for sealing during holiday periods. Others, such as hospital wards, are free only rarely and then for a very limited period. Others again, for example computer rooms, only become available at long intervals. Offices, crowded with tables and desks, require a great deal of preparatory work moving all the furniture before sealing can begin.

A seal with normal durability, such as an oleo-resinous seal, can be used if the floor is readily available. If it is not, the seal must be durable for the longest possible time, to reduce further inconvenience to a minimum. In the latter case, the best type of seal should be employed and a two-pot polyurethane seal is recommended—assuming the type of floor and circumstances are suitable.

Durability of Seal

In general, the longest possible durability is required. This is because the cost of sealing comprises about 90% labour costs and only 10% material costs. Any sealing operation is, therefore, costly with regard to labour, and the period between sealing operations should be extended as long as possible.

Whilst this is true in principle, there are exceptions. A less durable seal may be perfectly satisfactory on an open area, freely accessible and readily available when required. If the floor is due for alteration or demolition in the foreseeable future, an appropriate seal should be chosen to last the expected life of the floor.

Drying Time

Drying time can be a very important factor, particularly where floors must be back in use in the absolute minimum of time. With fast drying seals, for example, the two-pot polyurethanes, each coat dries in approximately 2 h. A floor can be sanded and sealed with three coats of this seal in one day.

If the drying time is not so important, as when sealing school floors during a holiday period, a longer drying seal may be perfectly satisfactory. Two coats of an oleo-resinous seal, for example, can be applied in 2 days, allowing overnight drying for each coat.

Quick drying can be an important factor in reducing the labour costs of a sealing operation. Not only is the time spent on the site shorter, but in many instances travelling time to and from the job is reduced. If a considerable distance has to be travelled to the floor being sealed, two or more visits may be necessary if the drying time is long and seal has to be allowed to dry overnight. If a quick drying seal is used, one visit only may be all that is required to complete the sealing operation.

Solvent Odour

Solvent odour can be important if it is pungent, or if part of the floor being sealed remains occupied, or people are occupying adjoining rooms. For this reason, whenever a seal with a comparatively strong odour is contemplated, for example a two-pot urea-formaldehyde seal, maximum ventilation must be allowed. If necessary, fans should be used to drive the solvent fumes away.

Application of oleo-resinous and water-based seals, with characteristically mild odours, should never present any odour problems.

Chemical Resistance

All seals develop resistance to chemical attack, however limited, and some do so more than others. Most, for example, will withstand water, oil and grease, although few will withstand attack by strong acids and alkalis.

If a seal is required for use on a floor liable to be affected by chemicals, enquiries should be made to ascertain which are resistant to the chemicals in question. This will narrow the choice of seal considerably.

It is generally recognized that the best possible resistance to chemicals is given by two-pot polyurethane seals and these have been used very successfully in laboratories, chemical warehouses, garages, hospitals, food factories and other areas subject to constant chemical attack.

Resistance to Yellowing on Ageing

This can be of importance if the seal is to be used on a light coloured floor. On a normal, medium-to-dark coloured wood floor, slight yellowing would not be noticeable. If, however, a very light coloured wood is to be sealed, one of the paler seals would be preferred.

More important, perhaps, is the yellowing which may take place if a seal is applied to a floor covering, such as linoleum. If a seal yellows on linoleum the appearance can deteriorate very rapidly, particularly if the linoleum is a light blue or turquoise colour, as this will assume a greenish appearance.

FACTORS AFFECTING THE CHOICE OF SEAL

Yellowing of films of water emulsion floor wax on white thermoplastic or vinyl tiles can become very unsightly. A water-based floor seal, based on acrylic polymer resins, will not yellow on ageing. If the floor is treated with an acrylic polymer seal and maintained with an acrylic emulsion floor wax, it will retain its white colour almost indefinitely.

Flammability

All solvent-based seals are flammable and all water-based seals are non-flammable. The degree of flammability of the solvent-based seals varies considerably. The oleo-resinous seals have high flash points, generally about 41°C (106°F) and are, therefore, relatively safe. Some seals based on nitrocellulose, on the other hand, have low flash points and are highly flammable.

Seals with low flash points must be stored away from boilers and must not be exposed to naked lights of any sort. Smoking should never be permitted when a solvent-based seal is being applied. Many solvents are heavier than air and can travel along the ground for some distance if lids are left off containers. A lighted match thrown down several yards from an uncovered container full of seal could cause a fire. It is recommended that notices forbidding smoking are placed at approaches to the area being sealed so that all are aware of a potential fire hazard.

Dried films of seal vary with regard to their flammability. Two-pot polyurethane seals, for example, are noted for their low flammability.

Slip Resistance

Slip resistance is always an important factor to consider. Nowhere is it more important than in hospitals and homes for elderly and infirm people. Whilst a fully slip resistant floor is desirable in most instances, it is absolutely essential in these cases.

Almost all seals reduce the slip hazard that is ever present on floors; a sealed floor is generally safer than an unsealed floor. The resistance of a seal to slip can be measured numerically in terms of coefficient of friction. The higher the coefficient the safer and more slip resistant is the seal.

It is a common fallacy to assume that because a floor looks glossy it must be slippery. In fact there is no relationship at all between gloss and slip, and many eggshell and matt seals are less non-slip than those with a high gloss. Despite the facts, however, many people associate in their minds gloss with slip and, therefore, feel safer on eggshell or matt surfaces. Where there is any possibility of gloss being associated with slip, in the interests of safety preference should be given to those seals drying with a subdued gloss, or even matt finish, rather than a high gloss.

Flexibility

Seals are generally assumed to be formulated so that they are flexible. Whilst this is indeed the case, some by their very nature are more flexible than others.

Urea-formaldehyde seals, for example, are generally not as flexible as some of the other types, whatever the formulation. Two-pot polyurethane seals are among the most flexible and are recommended where this quality is particularly required.

Cork and soft wood floors are easily marked by stiletto heels (see page 30). Care should, therefore, be taken to ensure that a flexible seal is used on a soft floor and not one that will become brittle and eventually flake off or puncture. Whilst flexibility is, in general, a minor factor, it assumes some importance when such floors are being considered.

Recoatability

Recoatability, the ease with which worn seal can be either touched-up or recoated, is important, particularly if it is intended to recoat at regular intervals.

Some seals, for example, oleo-resinous and synthetic rubber pigmented seals, can be readily touched-up and recoated as and when required. Others, particularly the two-pot seals, require a considerable amount of preparation. Even then, touching-up worn areas is not an easy task.

If skilled or experienced labour is not readily available and frequent touching-up or recoating is required, the simpler types of seal should be given serious consideration in preference to the more complicated plastic seals.

Removability

When sealing a floor is considered, it should always be recognized that any seal will eventually wear and that none will last for ever. At some time in the future, it is very likely that removal of the seal will be required.

If the seal is to be applied to a floor that can be sanded, wood, wood composition, cork and magnesite, for example, then there are no problems. If, however, it is intended to seal a floor that cannot be sanded, such as linoleum, thought must be given as to how the seal is to be removed when the time comes. Removal of seals from such floors is often difficult and laborious, particularly if chemical strippers have to be used.

Whilst solvent-based seals should only be applied to floors that can be sanded, this does not, of course, apply to water-based seals. These seals, which are, in general, less tough than the former, can be removed using an alkaline detergent to which a special solvent has been added, or with abrasive nylon mesh discs. These discs can also be used to remove some solvent-based seals from floors that cannot be sanded. Sanding is, however, a far more positive way of removing seal.

Subsequent Method of Maintenance

Having sealed a floor it is desirable to maintain it in good condition for as long as possible. Regular application of a floor wax, either solvent-based or water emulsion will extend the life of the seal. Consideration should always be given to the subsequent method of maintenance before the seal is chosen.

If the floor is readily accessible and can be easily maintained with a floor wax, it is not absolutely essential to use the most durable type of seal, as the wax will protect the seal from abrasion to some extent.

If, however, the floor is covered with furniture or other fittings, so making the use of a wax difficult, and it is intended only to sweep and damp mop using a solution of neutral detergent in water, the most durable type of seal should be used. In these circumstances, a two-pot polyurethane is recommended.

Advisability of Using a Priming Coat

All seals will only give of their best if they can key or adhere satisfactorily to the floor, otherwise they will quickly become detached and flake off.

Many seals are supplied in one grade, or quality, only. Clearly these seals lack the advantages of those that are supplied in two qualities, namely a priming and a finishing coat. The advantage of a priming coat is that it is specially formulated to penetrate the floor being treated thereby binding itself to it. The finishing coat forms a strong chemical bond with the priming coat and provides the hard, durable wearing surface.

By using a priming and finishing coat system better results are obtained than by using one quality alone. This is particularly the case on wood and similar floors, where hardness varies considerably from one to the next. Also, on concrete floors, the priming coat generally contains special alkali resistant materials, to combat the neutral alkali in the concrete. In both cases, the finishing coats provide an attractive appearance together with excellent abrasion resistant properties.

Cost, Initial and over an Extended Period

When considering cost, the initial cost of labour and materials for sealing the floor must be taken into account together with the total over a number of years. It may be necessary to re-apply a cheap floor seal several times over a period of a few years. Each sealing operation could prove a costly item in itself, particularly from the point of view of the amount of labour involved. If furniture needs to be moved before the floor can be sealed, the labour costs involved in removing and replacing the furniture must be included.

A top quality seal, although initially dearer, will probably prove to be cheaper over the same period of time because of its greater durability and the reduced labour costs incurred in re-sealing at less frequent intervals.

A consideration of material costs alone can prove very misleading. As stated earlier, they account for only about 10% of the total sealing costs. Greatest consideration, therefore, should be given to keeping labour costs at a minimum. In general, this can be achieved by using the best quality seal available.

These, then, are the main factors that should be considered when choosing a floor seal. Once all factors have been thoroughly investigated, the correct type of seal to use should become evident.

Some applications and properties of different floor seals are compared in *Table 2.1*.

Table 2.1. *Comparison of Floor Seals*

Floors	Clear Seals				
	Oleo-resinous	One-pot plastic		Two-pot plastic	
		Urea-formaldehyde	Poly-urethane	Urea-formaldehyde	Poly-urethane
Recommended for use on	Wood Wood composition Cork Magnesite	Wood Wood composition Cork Magnesite	Wood Wood composition Cork Magnesite	Wood Wood composition Cork Magnesite	Wood Wood composition Cork Magnesite
Can be used on			Concrete		Concrete Asphalt Linoleum

Table 2.1. Comparison of Floor Seals (cont'd)

Floors	Clear Seals				
	Oleo-resinous	One-pot plastic		Two-pot plastic	
		Urea-formaldehyde	Poly-urethane	Urea-formaldehyde	Poly-urethane
Must not be used on	Thermoplastic tiles PVC(vinyl) asbestos Flexible PVC Rubber Asphalt Terrazzo Marble Quarry tiles Stone Linoleum Concrete	Thermoplastic tiles PVC(vinyl) asbestos Flexible PVC Rubber Asphalt Terrazzo Marble Quarry tiles Stone Linoleum Concrete	Thermoplastic tiles PVC(vinyl) asbestos Flexible PVC Rubber Asphalt Terrazzo Marble Quarry tiles Stone Linoleum	Thermoplastic tiles PVC(vinyl) asbestos Flexible PVC Rubber Asphalt Terrazzo Marble Quarry tiles Stone Linoleum Concrete	Thermoplastic tiles PVC(vinyl) asbestos Flexible PVC Rubber Terrazzo Marble Quarry tiles Stone
Physical properties					
Colour	Medium	Very pale	Medium	Very pale	Pale
Solvent odour	Very mild	Moderate	Mild	Strong	Strong
Chemical resistance	Fair	Fair	Very good	Very good	Excellent
Durability	Good	Good	Very good	Very good	Excellent
Recoatability	Very easy	Fair	Fair	Requires care	Requires care
Removability	Easy	Fair	Fair	Difficult	Difficult
Shelf life	Very good	Poor	Very good	Base—very good Accelerator—very good	Base—very good Accelerator—poor
Drying time (hours per coat)	8–10	6–8	3–4	3–4	2
Minimum interval between coats (hours)	6	6	3–4	6	2
Maximum interval between coats (hours)	Indefinite	Indefinite	24	Indefinite	24
Hardening time before traffic allowed on floor (hours)	12–16 (overnight)	12–16 (overnight)	12–16 (overnight)	12–16 (overnight)	Minimum 4, preferably 12–16 (overnight)
Practical Data					
Recommended minimum number of coats	2–3	2–3	2–3	2–3	3
Recommended applicators	Roller Bonnet Brush Mop	Roller Bonnet Brush	Roller Bonnet Brush	Roller Bonnet Brush	Roller Brush

Table 2.1. Comparison of Floor Seals (cont'd)

Practical data (cont'd)	Clear Seals				
	Oleo-resinous	One-pot plastic		Two-pot plastic	
		Urea-formaldehyde	Poly-urethane	Urea-formaldehyde	Poly-urethane
Approximate coverage (ft^2/gal)	500–600	700–800	600–700	600–800	600–800
Brush cleaning solvent	White spirit	Special solvent	White spirit or special solvent	Special solvent	Special solvent
Type of floor wax for maintenance	Solvent wax (better) or water emulsion	Solvent wax (better) or water emulsion	Solvent wax (better) or water emulsion	Solvent wax (better) or water emulsion	Solvent wax (better) or water emulsion

Floors	Pigmented seals		Water-based seal	Silicate dressing
	One-pot synthetic rubber	Two-pot polyurethane		
Recommended for use on	Concrete Magnesite	Concrete Magnesite Asphalt	Thermoplastic tiles PVC(vinyl) asbestos Flexible PVC Rubber Asphalt Terrazzo Marble Linoleum	Concrete
Can be used on	Asphalt Stone	Stone	Magnesite Concrete Stone Quarry tiles	Some stone
Must not be used on	Thermoplastic tiles PVC(vinyl) asbestos Flexible PVC Rubber Terrazzo Marble Linoleum Quarry tiles Wood Wood composition Cork	Thermoplastic tiles PVC(vinyl) asbestos Flexible PVC Rubber Terrazzo Marble Linoleum Quarry tiles Wood Wood composition Cork	Wood Wood composition Cork	All other types floor
Physical Properties Colour	Coloured	Coloured	Very pale	Varies

Table 2.1. Comparison of Floor Seals (cont'd)

Physical properties (cont'd)	Pigmented seals		Water based seal	Silicate dressing
	One-pot synthetic rubber	Two-pot polyurethane		
Solvent odour	Mild	Moderate	None	None
Chemical resistance	Good	Excellent	Fair	Poor
Durability	Good	Excellent	Requires protecting	Fair
Recoatability	Very easy	Requires care	Very easy	Very easy
Removability	Easy	Difficult	Easy	Difficult
Shelf life	Very good	Base—very good Accelerator—poor	Very good	Very good
Drying time (hours per coat)	3–4	2	20–30 min	12–14
Minimum interval between coats (hours)	6	2	30 min (approx.)	12–14
Maximum interval between coats (hours)	Indefinite	24	Indefinite	Indefinite
Hardening time before traffic allowed on floor (hours)	12–16 (overnight)	Minimum 4 Preferably 12–16 (overnight)	Minimum 30 min, preferably 1 h	12–16 (overnight)
Practical data Recommended minimum number of coats	2	3	1–2	2–3
Recommended applicators	Roller Bonnet, brush Turk's head	Bonnet Brush Turk's head	Bonnet Mop	Watering can
Approximate coverage (ft²/gal)	300–500	500–600	3000–4000	300–350
Brush cleaning solvent	Special solvent	Special solvent	Water	Water
Type of floor wax for maintenance	Water emulsion	Solvent wax (better) or water emulsion	Water emulsion	None (or water emulsion)

3

PREPARATION OF FLOORS FOR SEALING AND APPLICATION OF SEAL

PREPARATION OF FLOORS FOR SEALING

IT is absolutely essential that the floor is properly prepared if the best possible results are to be obtained from any sealing operation. Many floors are subjected to relatively heavy traffic over a short period. If there is any imperfection in the adhesion between the seal and floor it will quickly become evident and the seal will wear off.

The life of any seal depends very largely on how well it adheres to the floor. Adhesion depends almost entirely on the manner and thoroughness in which the floor is prepared for sealing. A number of factors determine the way in which this is done. As technology in the floor maintenance field progresses, so floor maintenance materials become more complex and correct preparation even more important. Poor preparation can reduce the life of a seal by months, or even years. In the worst possible circumstances all seal may have to be removed and the surface re-prepared.

This chapter deals with the methods by which each type of floor should be prepared to receive seal. 'Seal', in this context, is again taken to mean 'a semi-permanent material which, when applied to a floor, will prevent the entry of dirt and stains, liquids and foreign matter'. This definition includes solvent- and water-based materials. Details of procedures regarding the preparation of floors to receive both types of seal are, therefore, included.

In general, floors can be grouped into the following categories:

(a) Floors that should be sealed with a solvent-based seal.

Wood   Wood composition
Cork

(b) Floors that can be sealed with a solvent-based seal if required.

Linoleum   Magnesite
Concrete   Asphalt

PREPARATION OF FLOORS FOR SEALING

(c) Floors that should be sealed with a water-based seal.

        Thermoplastic tiles   Linoleum
        PVC(vinyl) asbestos   Rubber
        Flexible PVC   Marble
        Terrazzo

(d) Floors that should not be sealed.

        Quarry tiles   Stone

Let us now consider each group in turn.

(a) **Floors that Should be Sealed with a Solvent-based Seal**

Unsealed wood, wood composition and cork floors are porous and will absorb water and dirt. As a result the floors are difficult to maintain and rapidly become unsightly.

A new floor should be sanded to remove any imperfections and finished off with a fine grade paper. The floor should then be vacuumed or damp mopped to remove dust; brushing will merely raise the dust to allow it to settle again later.

If the new wood floor consists of wood block or parquet, extra precautions should be taken, if a plastic seal is to be used, to prevent the possibility of blocks becoming stuck together with seal, with subsequent ' rafting ' leaving unsightly gaps in the floor. ' Rafting ' is the movement of a number of blocks simultaneously, causing a crack to appear in the floor. It can be caused if the blocks are subjected to large changes in moisture content, making them shrink excessively, whilst they are tightly bonded together by means of seal.

A new wood block or parquet floor should first be treated with a thin application of a free-flowing liquid solvent wax over the whole area, so that it penetrates between the blocks. The wax acts as a lubricant and enables each block to move a fraction individually on shrinking, so eliminating the possibility of ' rafting '. When dry, the floor should be sanded to remove all traces of wax from the surface before seal is applied.

Ideally, old floors should always be sanded prior to sealing. This is to remove all dirt and previous surface treatments which may prevent the new seal adhering properly. A seal is formulated especially for application direct to the floor itself and, if it is applied on top of another material, the adhesion and hence durability of the seal may be impaired. Also, it should be recognized that different types of seal may be incompatible; for example, the strong solvents

PREPARATION OF FLOORS FOR SEALING; APPLICATION OF SEAL

in some plastic seals will soften and lift oleo-resinous seals. Conversely, the weak solvents in oleo-resinous seals will not be sufficiently strong to soften some plastic seals, with the result that a new application of an oleo-resinous seal will not adhere to the plastic seal.

If, however, sanding is not possible for any reason, the way in which the surface should be prepared will depend upon the methods by which the floor has been maintained.

Where the floor has been treated with button polish or has been only lightly sealed, it can be prepared by stripping the surface using an abrasive nylon mesh disc, grit 80 is recommended, under an electric polishing/scrubbing machine. Abrasive nylon mesh discs give results similar to those obtained by a very light sanding operation. After stripping, all dust should be removed from the surface with mopping equipment and the floor allowed to dry before sealing.

If the floor has been maintained with a solvent-based wax, either paste or liquid, it is absolutely essential for all traces of wax to be removed, because no seal will adhere to wax. Removal can be carried out using a solvent-based detergent wax remover. The material should be applied to the floor, and allowed to soak into the wax layer for a few minutes, then scrubbed using a metal or nylon web pad under an electric polishing/scrubbing machine. The loosened wax can then be removed using mopping equipment. It is important to change the water frequently to ensure that the wax is completely removed from the floor and not just redeposited elsewhere.

It is always recommended that a test should be carried out on a floor stripped of wax to ensure that all wax has, in fact, been removed. The test should be conducted on an area of about 1 or 2 yd^2, preferably on the area of floor cleaned last, because if wax is still present on the floor it is most likely to be found here.

To test, the seal should be applied and the drying time and hardness checked. If the seal dries hard in the specified time it is reasonable to assume that all wax has been removed and the floor can be sealed. If the seal does not dry hard in the specified time, or dries patchily, it is possible that wax has not been properly removed and the de-waxing operation should be repeated, followed by a further test.

If the floor has been maintained with a water emulsion floor wax this should be removed using an alkaline cleanser. Once the floor is clean it should be thoroughly rinsed to ensure that no alkali remains on the surface. As an extra precaution it is advisable to add a little vinegar to the rinsing water so that any remaining

PREPARATION OF FLOORS FOR SEALING

alkali is neutralized. Failure to rinse thoroughly may result in the seal being affected by the alkaline detergent with a consequent darkening in colour.

Floors that have previously been treated with an oleo-resinous seal should be cleaned and any wax used to maintain the floor removed. The old seal should, ideally, be roughened up using coarse grade metal fibre or nylon web pads, to provide a mechanical key for the new application. It is always preferable to finally mop over using white spirit to remove all traces of water and very slightly soften the old seal, so that it is in a receptive condition for the new seal.

If the floor has been previously sealed with a plastic seal, the procedure is similar to that described above. Because plastic seals are harder than oleo-resinous seals, greater attention should be given to the roughening up process, to ensure that all traces of gloss are removed from the surface. Finally, the floor should be mopped using the appropriate solvent to again slightly soften the old seal. The new seal should be applied as soon as the solvent has evaporated. Large areas of floor should be treated in sections. Sanding is, of course, essential if the new seal is not compatible with the old.

Further details regarding the compatability of seals are given in the section on ' Resealing ' (page 99).

(b) Floors that Can be Sealed with a Solvent-based Seal if Required

Linoleum

As a general rule it is advisable to maintain linoleum with a floor wax. It can, however, be sealed with either a water- or solvent-based seal, if required. A water-based seal is preferred and can be used with particular advantage on porous linoleum, particularly the thick, ' battleship ' grades. The use of a water-based seal will be considered first.

New linoleum, or a floor which has not previously been treated, should be cleaned with a solution of neutral detergent in water, using a fine or medium grade metal fibre or nylon web pad, if necessary. When clean, the floor should be rinsed well and allowed to dry before seal is applied.

If the floor has been maintained with a solvent-based wax, all wax must be removed using a solvent-based detergent wax remover. The material should be applied to the floor, allowed to soak for a few minutes, then scrubbed using a metal fibre or nylon web pad under an electric polishing/scrubbing machine. The loosened wax

PREPARATION OF FLOORS FOR SEALING; APPLICATION OF SEAL

can then be removed using mopping equipment. It is important to change the water frequently to ensure that the wax is completely removed from the floor and not just redeposited elsewhere. After cleaning, the floor should be well rinsed and allowed to dry.

A floor previously treated with a water emulsion wax should be stripped of wax using an alkaline detergent solution in water which should be applied to the floor and allowed to soak for a few minutes. Soaking is important, otherwise the detergent does not have time to act on the old film of floor wax and removal is rendered more difficult. After soaking, the floor should be scrubbed using a metal fibre or nylon web pad under an electric polishing/scrubbing machine. The loosened floor wax can then be removed with mopping equipment.

When an alkaline detergent has been used, it is important that no residue remains after cleaning, as this would prevent adhesion of seal. The floor should, therefore, be rinsed thoroughly. To assist in neutralizing any alkali which may remain, it is again advisable to add a little vinegar to the rinsing water. After rinsing thoroughly, the floor should be allowed to dry.

Where a floor has previously been sealed with a water-based seal, any water emulsion floor wax used as a surface dressing must be removed with an alkaline detergent, using the method described above, before further application of seal is made.

If the floor has been sealed with a floor paint, oleo-resinous or plastic seal, re-sealing with a water-based seal should not be attempted because of possible difficulties with adhesion and also because the results obtained are likely to be patchy and unsatisfactory.

Reference has been made to the sealing of linoleum with two-pot polyurethane seals (page 53). In general, treating linoleum with a solvent-based seal is not recommended, because of the difficulties of later removal. If, however, such a seal is considered necessary, perhaps to lengthen the life of the linoleum prior to replacing it or to make it chemically resistant, a two-pot polyurethane, and no other type, should be used. It must be emphasized that instances where linoleum should be sealed in this manner are few and far between.

Any wax present must be removed by using the appropriate material, as described above. Linoleum generally has a surface dressing, applied during the manufacturing process, which must also be removed before sealing with a two-pot polyurethane seal. To remove the manufacturer's dressing, treat a small section at a time with the appropriate solvent, allow it to soak for a few minutes

PREPARATION OF FLOORS FOR SEALING

and scrub with a metal fibre or nylon web pad under an electric polishing/scrubbing machine. Follow this operation immediately by washing with a solvent-based detergent wax remover. Remove the loosened dressing with mopping equipment, rinse the floor thoroughly and allow to dry.

The result will be a matt appearance. If any glossy patches appear, the scrubbing process should be repeated until the linoleum assumes a uniform matt appearance throughout. Once all the manufacturer's dressing has been removed the floor can be sealed.

If the linoleum has previously been sealed with a solvent-based seal, it should be cleaned and the remainder of the seal roughened up using coarse grade metal fibre, nylon web pads or grit 120 abrasive nylon mesh discs under an electric polishing/scrubbing machine. The floor should be washed to remove all dirt and allowed to dry, and then mopped over using the appropriate solvent which will slightly soften the old seal and provide a key for the new coat. Seal should be applied immediately after the solvent has evaporated.

Magnesite

Magnesite floors generally have a wood filler and because of this behave similarly to wood. It is normal for flooring contractors to apply a seal to magnesite floors when they are laid. When the surface shows signs of wear, it is advisable to again apply a seal to maintain the good appearance and to prevent dusting and excessive wear. Magnesite floors are generally coloured. If loss of colour is evident, due to wear, it can be restored either by using a coloured emulsion wax on a clear seal, or by using a pigmented seal.

If it is desired to seal a new floor, the surface dressing must be removed first, using an alkaline liquid detergent in water. After scrubbing, the floor should be rinsed well, with a little vinegar in water and allowed to dry. A clear or pigmented seal can then be applied.

A floor that has been maintained with a solvent-based wax, water emulsion floor wax, or seal should be prepared in the manner described in the section dealing with linoleum, above. It is, however, important to note that magnesite is generally very absorbent. Only the minimum amount of water should, therefore, be used and adequate time allowed for the floor to dry hard before seal is applied. This is particularly important if a two-pot polyurethane seal is to be used, as the accelerator is very sensitive to moisture. If the floor is at all damp, the accelerator will react with the moisture in the floor rather than with the base, causing unsightly bubbles to appear and remain in the film of seal.

PREPARATION OF FLOORS FOR SEALING; APPLICATION OF SEAL

A water-based seal can also be applied to magnesite, if desired, in which case a coloured seal is preferred. The floor should be stripped of all wax by the methods outlined above. One or two coats should be applied, followed by maintenance with a water emulsion floor wax.

Concrete

Concrete floors, even those of the granolithic type, have a tendency to dust. Sealing is, therefore, always desirable and often essential, both to eliminate dusting and to improve the appearance of the floor. New concrete must be allowed to harden for at least 14 days and there must be no evidence of water in the pores. To achieve maximum adhesion and therefore durability, the concrete should first be etched with etching crystals. This is particularly important if the surface is steel trowel finished or very close grained.

The crystals are slightly acidic and attack the surface of the concrete, opening it fractionally and increasing its porosity. After using the crystals, the treated area must be well rinsed several times to remove all traces of the crystals and to leave the floor in a neutral condition. The floor should then be allowed to dry before it is sealed. If seal is applied too soon, water may be trapped under it in the concrete and later cause the seal to blister.

If the concrete is not new, it must be swept and cleaned before seal is applied. In the unlikely event of a floor wax having been used, this must be removed before sealing.

Where a floor has been sealed with a rubber-based floor paint, it should be thoroughly cleaned with an alkaline detergent, rinsed well and allowed to dry before resealing.

If a plastic seal, for example a pigmented two-pot polyurethane, has been used, the surface should be roughened up with coarse grade metal fibre or nylon web pad and washed to remove all dirt. It should then be mopped over using an approved solvent which will slightly soften the old seal and provide a key for the new application.

Concrete that has been treated with a silicate dressing should be swept to remove all dirt. If any wax has been applied to the floor that also must be removed before a further coat of silicate dressing is applied. Once clean and dry, a coat can be applied in the normal way.

Asphalt

Asphalt floors, which are generally laid in dark colours, principally red, brown and black, tend to lose their original colour under constant traffic and may appear faded.

PREPARATION OF FLOORS FOR SEALING

Whilst water-based seals are perfectly safe and are widely used on such floors, ordinary seals, based on solvents, cannot be used as the solvents tend to soften the floor. However, special seals, based on polyurethane or synthetic rubber, can be used satisfactorily, provided certain precautions are taken.

New asphalt floors generally have a surface film. Before applying any sort of seal it is advisable to first remove this film by scrubbing with an alkaline detergent solution in water. The floor should then be well rinsed and allowed to dry. If the surface film is not removed, the floor seal will 'crawl' (like water on grease), and dry with a spotty, uneven finish.

If a water-based seal is to be applied, a coloured material to match the floor should be used. The floor should then be maintained with a coloured emulsion floor wax.

Where a polyurethane seal is to be used, either clear or pigmented, a test area should be treated first to ensure that the floor is properly prepared and that the seal will adhere to the asphalt floor. These floors vary widely in composition, some consisting of hard and others of soft materials. A test area will, therefore, ascertain whether the seal will soften and ruin the floors or penetrate just sufficiently to obtain satisfactory adhesion.

The test should be carried out in one of the main traffic lanes, preferably about 4 yd^2. A small test area, alongside a wall away from traffic, is of no value whatsoever and may even give misleading results. The test should be carried out over a period of about 4 weeks, to allow the seal to harden properly and to be subjected to traffic. If the seal adheres satisfactorily after this period and there is no sign of wear, the floor is suitable for sealing. If, however, the seal blisters, shows signs of lack of adhesion or wears, then sealing should not be attempted with a polyurethane seal.

A pigmented synthetic rubber seal can be applied satisfactorily after removing the surface film from the asphalt floor. Again, if possible, a test area should be treated to ensure that the results meet with approval.

Old asphalt floors, previously treated with a water emulsion floor wax, should be scrubbed using an alkaline detergent in water to remove all traces of the floor wax. In the unlikely event of a solvent-based wax having been used on asphalt, it should also be removed with the detergent. Several scrubbings, using metal fibre or nylon web pads under an electric polishing/scrubbing machine, may be required to remove all traces of wax. Solvent-based detergent wax removers must not be used as the solvent component will soften and perhaps damage the asphalt. If the asphalt floor has

PREPARATION OF FLOORS FOR SEALING; APPLICATION OF SEAL

previously been sealed with a pigmented synthetic rubber or polyurethane seal, the surface should be roughened up using coarse grade metal fibre or nylon web pads under an electric polishing/scrubbing machine and washed to remove all dirt. It should then be mopped over with an approved solvent. Seal should be applied immediately after the solvent has evaporated.

(c) Floors that Should be Sealed with a Water-based Seal

The floors discussed below should not, in general, be sealed with solvent-based seals. Reasons vary and each floor, or group of floors, will be considered in turn. The majority can, however, be sealed very satisfactorily with a water-based seal and preparation for this will be detailed.

Linoleum

The treatment of linoleum with a water-based seal has been discussed on page 91 and readers should refer to this page.

Thermoplastic Tiles, PVC(vinyl) Asbestos and Flexible PVC

Solvents in solvent-based seals are liable to soften and damage both PVC(vinyl) asbestos and thermoplastic tile floors, and, since the pigments used in these floor coverings are generally soluble in solvent, often cause the colours to ' bleed ' and run into each other. There is, unfortunately, no method of putting right a floor covering in which this has happened, other than re-laying the affected area.

There are many grades of flexible PVC, and although a few are resistant to solvents, the majority are not. It is, therefore, advisable not to risk ruining the floor by applying solvent-based materials but to use a safe, water-based seal, if sealing is required.

New floors should not be washed for a period of 14 days to allow the adhesive to dry thoroughly. They should then be cleaned using an alkaline detergent solution, in conjunction with metal fibre or nylon web pads, followed by thorough rinsing. It is always advisable to add a little vinegar to the rinsing water to neutralize any alkaline residue that may be present on the floor. The floors should then be allowed to dry before a water-based seal is applied.

Old floors which have been maintained with a water emulsion floor wax should be stripped of wax using an alkaline detergent, as above. Seal can be applied as soon as the floor is dry.

Where a solvent-based wax has been used on these floors, all traces of it must be removed before seal is applied, using, in the

PREPARATION OF FLOORS FOR SEALING

case of flexible PVC, a solvent-based detergent wax remover. Whilst this operation can be carried out satisfactorily on most grades of flexible PVC, it is always advisable to test a square yard or so in one corner first to make certain the coloured pigments are fast to solvents. If PVC(vinyl) asbestos or thermoplastic tiles have been treated with solvent-based wax, the floor may be showing some signs of deterioration due to the action of the solvent. Solvent-based detergent wax removers cannot be used on these floors, because of the quantity of free solvent present which could further damage the surface being treated. An alkaline detergent should, therefore, be used. Although it is considerably more difficult to strip a solvent wax by this method, there is no alternative if the floor is not to be harmed. The procedure has been described above, thorough rinsing being essential for success.

Floors with a solvent-based seal should be examined carefully before any recommendations are made and each should be treated on its own merits. Whilst it is wrong to apply a solvent-based seal to PVC(vinyl) asbestos or thermoplastic tiles, because of the damaging effects of solvents on these surfaces, many floors have, in fact, been sealed with these materials. The floors may remain in an acceptable condition until the seal shows signs of wear, when re-sealing problems arise. Some seals have turned light coloured tiles to a dirty brownish-yellow shade in a relatively short period. In these cases, removal of seal should be carried out with abrasive nylon mesh discs under an electric polishing/scrubbing machine. Grit 120 discs, or finer, should be used. Coarser grit discs are not recommended as they may scour the floor surface and leave marks in the floor. After using the discs, all dust should be removed and one or two coats of water-based seal applied.

Rubber

Rubber floors can be maintained satisfactorily with a water emulsion floor wax. If, however, sealing is required then only a water-based seal should be used. This is because solvents soften rubber and may cause it to swell.

New rubber floors can be treated with a water-based seal as soon as the adhesive has dried. When cleaning prior to sealing, a neutral detergent can be used if dirt and soilage are light, and an alkaline detergent if the floor is very dirty. It is essential that only the minimum amount of water be used, to prevent it from creeping between the tiles or joints of the floor and affecting the adhesion of the rubber to the sub-floor.

PREPARATION OF FLOORS FOR SEALING; APPLICATION OF SEAL

Old floors which have not been maintained properly will usually be found to have a rough, oxidized surface, due to the action of air and sunlight. Before sealing with a water-based seal the oxidized surface must be removed. Removal can be carried out using fine grade metal fibre or nylon web pads under an electric polishing/scrubbing machine. If the oxidized surface is bad, a fine grade abrasive nylon mesh may be required. Grit 120 or finer discs should be used. After removing the oxidized surface layer, two or three buffings may be necessary using fine grade metal fibre or nylon web pads, to produce a surface suitable for sealing. The floor should then be swept or vacuumed and damp mopped to pick up any small particles that may remain. Sealing should be carried out as soon as the floor is dry.

Old floors which have been previously waxed with a water emulsion floor wax should be thoroughly stripped using an alkaline detergent, as described in the section on PVC(vinyl) asbestos, flexible PVC and thermoplastic tiles (page 96).

Floors with other than water-based seals should be treated individually, on their merits. In many cases the floors should not have been sealed in the first place and the best thing that can be done is to effect some form of repair. Rubber floors cannot be sanded, but fine grit abrasive nylon mesh discs may, in some instances, remove old seal. This done, the floor should be swept and damp mopped to remove all bits, allowed to dry and treated with water-based seal.

Terrazzo and Marble

Terrazzo and marble will be considered together because of the similarity of their composition. Both consist of marble, the terrazzo being composed of small irregular pieces set in a matrix of cement.

Solvent-based seals should not be used on these floors because of the difficulties of obtaining adhesion and because of possible yellowing, resulting in discolouration of the floor. Water-based seals are, however, recommended for use on both types of floor, to fill the open pores and provide a surface for future maintenance.

New floors should be prepared by cleaning with a solution of a neutral detergent in water. The floor should be allowed to dry before water-based seal is applied.

Old floors previously maintained with a water emulsion floor wax should be stripped, using an alkaline detergent, in the manner described on page 17. A water-based seal can be applied as soon as the floor is clean and dry. If the floor has been maintained with a solvent-based wax, all traces of this should be removed using a solvent-based detergent wax remover. The floor should be thoroughly rinsed and allowed to dry before applying seal. Floors which have been sealed with a solvent-based seal must be considered individually before sealing is attempted.

(d) Floors that Should not be Sealed
Quarry Tiles and Stone

Quarry tiles and stone surfaces are very similar in composition and properties and will be considered together. Neither should be sealed with a solvent-based seal as it is extremely difficult to effect penetration and obtain adhesion. Seals can be applied and will dry satisfactorily, but after a short time they detach from the surface and wear off.

Water-based seals generally offer little advantage except on porous stone floors. If the stone is particularly porous a water-based seal will fill the gaps and prevent dirt from penetrating, so making maintenance easier. This type of floor is, however, exceptional and, in general, sealing is not recommended.

RE-SEALING OF FLOORS

Reference has been made in the previous section to the difficulties involved when a sealed floor is re-sealed at a later date, due to the wide variation in composition of seals. After a long period, some recoat easily whilst others develop a very pronounced chemical resistance and will resist a fresh application, so that it does not adhere and flakes off under traffic.

If it is intended to re-seal a sealed floor, certain precautions must be taken. It is not sufficient merely to clean the floor and apply the chosen seal, otherwise this may wear off after only a few weeks.

First, it is essential that the type of seal already on the floor be known, so that the new seal can be chosen. In general, this should be of the same type as the old one so that the best possible key can be obtained. For example, if the floor already has an oleo-resinous seal, the new seal should be of the same type; if the present seal is a two-pot polyurethane, then this should be used. The mistake is sometimes made of applying the latter on top of an oleo-resinous seal. If this is done, the strong solvent in it will soften and lift the oleo-resinous seal, giving a very unsatisfactory result.

On the other hand, if an oleo-resinous seal is applied on top of a two-pot polyurethane seal it will not make any impression on it and

PREPARATION OF FLOORS FOR SEALING; APPLICATION OF SEAL

will fail to key to the surface. The slighest traffic will cause the oleo-resinous seal to flake off and detach from the floor.

Care must, therefore, be taken to ensure the new seal is compatible with the old. If it is not, all old seal must be removed, by sanding, by the use of abrasive nylon mesh discs or with a chemical stripper. Chemical strippers are generally based on methylene chloride, which has a strong odour. For this reason they should only be used on small, well-ventilated areas.

If the origin or type of the old seal is not known, and the floor is not to be sanded or stripped with an abrasive nylon mesh disc, a test should be carried out on an area of a few square yards. The longer the test area is subjected to traffic, the better. If adhesion is satisfactory after about 4 weeks it can be assumed that the seals are compatible. If considerable wear of the new seal has taken place it will be evident that the seals are incompatible, and the floor should be sanded before being sealed.

Table 3.1, showing compatibility of the main types of seals, has been compiled from both theoretical and practical considerations. It is intended to give an indication, and only an indication, of the possibilities of applying a new seal onto an old seal. There will, of course, be exceptions to the rule, but the table is substantially correct under normal circumstances.

Table 3.1. Compatibility of Seals

Old Seal \ New Seal	Two-pot polyurethane	Two-pot urea-formaldehyde	One-pot polyurethane (oil-modified)	One-pot urea-formaldehyde	Oleo-resinous	Synthetic rubber seal	Silicate dressing
Two-pot polyurethane	C	C	C	C	X	C	X
Two-pot urea-formaldehyde	C	C	C	C	X	C	X
One-pot polyurethane (oil-modified)	C	C	C	C	S	C	X
One-pot urea-formaldehyde	C	C	C	C	C	C	X
Oleo-resinous	X	X	S	X	S	S	X
Synthetic rubber seal	X	X	S	X	S	S	X
Silicate dressing*	C	C	C	C	C	C	S

S—New seal can be satisfactorily applied on to the old seal, providing the floor is clean, free from all trace wax, oil and grease, and dry.

C—New seal can be applied on the old seal provided extra care is taken. Floor must be (a) free from all tra of wax, oil and grease, and dry; (b) roughened up so that no gloss remains on the surface. (This gi the new seal a mechanical key to the old); (c) mopped with appropriate solvent immediately prio application of the new seal. (This gives the new seal a chemical key to the old.)

X—New seal cannot be satisfactorily applied to old seal under normal circumstances.

* If the type of seal is to be changed, all surface silicate dressing must be removed with etching crys before new seal is applied.

APPLICATION OF SEAL

If the old seal is of a type of material other than those shown, it is always advisable to check with the manufacturers of the new seal that they are compatible before any is applied.

APPLICATION OF SEAL

The application of seal is well within the capabilities of the handyman. Like every job of this type, particular emphasis must be given to the initial preparation. If the preparatory work is not carried out properly the best results will not be obtained. A little extra care is well worth while when applying the more sophisticated two-pot seals, to ensure that the highest standards of which they are capable are obtained.

Ideally, sealing should be carried out at average room temperature, which in the United Kingdom is 18·3°C (65°F). If the temperature is very much colder it will probably slow down the drying rate and it would be advisable, in this case, to introduce some form of heating to raise the temperature in the area being sealed.

Good ventilation is also necessary. If ventilation is poor, solvent becomes trapped and cannot escape with the result that solvent vapour remains in contact with the film of seal and retards drying. If ventilation by normal means is difficult, fans should be provided to assist in the removal of solvents. This is particularly important in basements and other poorly ventilated areas.

Applicators currently in use include brushes, mops, lambswool bonnets, roller applicators and turk's head brushes; in addition, watering cans are used to apply silicate dressings, but these will not be considered further in this section.

Brushes

Brushes are very useful for sealing small areas or the edges of areas which cannot be satisfactorily reached using another type of applicator. 2 in brushes are used very successfully to seal areas in corners or under heating pipes and indentations or faults in the floor itself, which cannot be properly sealed using an applicator.

Large areas should never be attempted with brushes as their use is slow and laborious compared with other methods.

Mops

Mops are very satisfactory for applying oleo-resinous seal. Cotton cord mops should be used and it is important that they are of good quality, and do not shed lint or bits of cotton into the seal.

The normal method is to pour about a cupful of seal onto the floor and spread over an area of a few square yards. This process

PREPARATION OF FLOORS FOR SEALING; APPLICATION OF SEAL

is repeated until the whole floor is covered. Care must be taken not to apply an excess of seal. Should this happen the surplus should be spread on to the next area to be treated.

Mops can also be used to apply water-based seal, and large areas can be covered quickly, particularly with the fringe type.

Lambswool Bonnets

These are used to apply seal to large areas. They are generally attached to the stock of an old broom, from which the bristles have been removed, or to the end of a squeeze-applicator.

Seal is poured onto the floor and spread lightly and evenly over the surface. Long, even strokes should be used and the applicator lifted as little as possible to avoid causing air bubbles. The stroke should be finished off toward the operator.

Whilst lambswool bonnets are very satisfactory for applying most seals, including the water-based type, when applying two-pot plastic seals they have a tendency to leave a line of bubbles where the applicator is removed from the floor. If the seal is quick drying, the bubbles may not have time to rise to the surface and burst before the surface is dry, with the result that they are trapped in the seal, giving the floor an unsatisfactory appearance.

Lambswool bonnets are not, therefore, recommended for applying two-pot seals, for which roller applicators are preferred.

Roller Applicators

These are made of mohair, wound convolutely onto the applicator. The effect is to force seal into the floor in both vertical and horizontal directions, so giving the best possible penetration.

They are supplied with a metal tray and grill. Seal is poured into the tray and the roller is worked in it until all air trapped in the mohair is replaced by seal. Surplus seal is then removed by passing the roller over the grill before applying it to the floor.

Roller applicators are used for sealing large areas and are generally about 10% quicker than lambswool bonnets. In addition to oleo-resinous seals, they are ideal for applying two-pot plastic materials, as when used properly there is no tendency to leave bubbles in the seal. This is because the applicator is saturated with seal and no air is present on the working surface. Also, the roller can be ' rolled ' off the surface gradually, rather than lifted, at the end of a stroke.

Roller applicators cannot be used for applying water-based seal, as this has too low a viscosity. They are, however, ideal for solvent-based seals.

APPLICATION OF SEAL

Turk's Head Brushes

Turk's head brushes are used to apply seals, particularly the pigmented types, to concrete floors.

Most such floors are uneven and pitted. If a roller applicator is used, it only applies seal to the raised portions of concrete, missing the hollows altogether. The recommended procedure, therefore, is to apply seal thinly and evenly to an area of a few square yards with a roller applicator or lambswool bonnet, then use a turk's head brush to spread the seal already on the floor to the lower, pitted, areas. When used in this way all the concrete floor is sealed and no gaps remain.

Some practical tips, which could make all the difference between a poor and an excellent result, follow.

It is always advisable to fully prepare the floor before getting the seal ready. This is particularly the case when two-pot seals are used, as the pot-life is limited. Base and accelerator should be mixed immediately prior to use and not some hours beforehand, otherwise the seal may thicken and become unusable.

When using two-pot seals, the mixed material should be allowed to stand for about 5 min before being used. This gives time for the material to mix properly and for the air bubbles introduced during the mixing process to surface and burst. If a two-pot seal is applied immediately after it has been mixed, an excessive amount of bubbles may appear on the floor. Whilst many will burst, some may become trapped and spoil the finished result.

In cold weather many seals tend to thicken a little and do not flow as well as they normally do, resulting in too thick a film and loss of coverage. One solution is to pre-heat the cold seal by placing the container against hot water pipes or in a hot room for some time prior to use. This will lower the viscosity and result in better flow, a thinner, more even film and normal coverage.

When planning the sealing operation, extra attention should be given to entrances and the main traffic lanes. It is always recommended that these should be given one or two coats of seal first, before the whole area is sealed. This will help to ensure that the seal wears evenly and that the areas subjected to most traffic have the greatest protection.

In gymnasia or other areas where white or coloured lines are often used, the lines should be applied after the first coat. Further coats of seal protect the lines from wear, with the result that they last much longer than if they had been painted on the surface. It is, therefore, essential that the line marking material is compatible with the seal, or one may lift the other.

It is always advisable to apply seal in daylight because it is often very difficult to see properly in artificial light and some areas may be missed altogether. If, however, sealing must be carried out under artificial light, the position of the light switches should be noted before sealing begins, so that the lights can be turned out when the job is finished. Where more than one coat is to be applied, a gentle flatting down of each coat will help to show where seal is present when applying the next coat.

When applying seal the aim should be to produce thin, even coats. Two thin coats are always better than one thick one. If coats are too thick, bubbles may not be able to escape and may be trapped in the film. Thick films of polyurethane seal are milky in appearance and the floor may assume a whitish colour. If an oleo-resinous seal is applied too thickly, it tends to form 'puddles', which may require several days to dry hard. If traffic is allowed over the floor in the meantime, the top of each puddle will quickly wear off exposing the undried, semi-liquid material underneath. This will stick to the soles of shoes leaving an unsealed area on the floor, obviously undesirable apart from the inconvenience to the people concerned.

As soon as the sealing operation has been completed it is essential that all equipment is cleaned immediately. It is not sufficient merely to place brushes or applicators in solvent or water. In the case of the two-pot plastic seals particularly, the hardening reaction between base and accelerator will continue regardless of conditions.

Applicators should be cleaned in the appropriate solvent, if necessary several times. It is then advisable to clean them in soap and water, rinse well and allow to dry.

MAINTENANCE OF SEALED FLOORS

Having obtained a pleasing finish by using a seal it is desirable to maintain this in the best condition for as long as possible. Regular application of a good quality floor wax will help to ensure that the natural attractiveness of the floor is maintained and that the life of the seal is considerably extended. Frequency of application of wax will, of course, depend upon the type and volume of traffic.

Whether a solvent-based or water emulsion floor wax is used is largely a matter of personal preference; both give excellent results. Whilst there is little to choose between the two types, solvent waxes are generally preferred on sealed wood, wood composition and cork

DURABILITY OF SEALS

floors, since, if the seal wears through, it is preferable that solvent, rather than water, comes into contact with the floor itself.

DURABILITY OF SEALS

A considerable amount of research and investigation has been carried out to determine how the durability of floor seals is extended by regular maintenance with a floor wax.

It was found, as was expected, that the durability of each seal depends largely on two factors. The type and volume of traffic and maintenance after sealing. Areas were, therefore, divided into those carrying 'light' and 'heavy' traffic, 'with' and 'without' maintenance. 'Light' traffic is found, for example, in gymnasia and office areas; 'heavy' traffic, in entrance halls, busy corridors and factory work areas, etc. By 'with maintenance' is meant periodic application of a floor wax; by 'without maintenance', no maintenance other than normal and damp mopping.

It is not normal practice to maintain pigmented seals with a floor wax and no observations were made on waxed pigmented seals.

It is stressed that the results given below are averages obtained from a great many areas. Individual results in each group varied widely, depending upon the factors mentioned above. Some floor seals were in excellent condition after double the average number of years; others, probably due to poor preparation of the floor, wore off after a few months. However, under average conditions, it is reasonable to expect that durability will be in the region of the figures shown in the table below, which are intended for use as a guide only.

Table 3.2. Comparative Durability of Seals

Type of Seal	Traffic	Average Durability (Years)	
		Without Maintenance	Maintained with wax
Oleo-resinous and One-pot urea-formaldehyde	Light Heavy	$1\frac{1}{2}$–2 1	2–3 $1\frac{1}{2}$–2
One-pot polyurethane (Oil-modified)	Light Heavy	$1\frac{1}{2}$–$2\frac{1}{2}$ $1\frac{1}{2}$	$2\frac{1}{2}$–$3\frac{1}{2}$ $1\frac{1}{2}$–$2\frac{1}{2}$
Two-pot polyurethane (Clear)	Light Heavy	3–4+ $1\frac{1}{2}$–2+	4–5+ 2–3+
Two-pot polyurethane (Pigmented)	Light Heavy	2–3+ $1\frac{1}{2}$–2+	— —
Synthetic rubber pigmented seal	Light Heavy	$1\frac{1}{2}$–2 1	— —

4

FLOOR WAXES

INTRODUCTION

THE glossy appearance of a polished surface is due to the reflection of rays of light striking the surface. The polish itself may consist of wax, made smooth by buffing, or a number of ingredients having inherent ' dry-bright ' qualities.

That such an appearance is desirable is evidenced by the amount of time and effort devoted to achieving it. The provision and maintenance of a polished surface plays a major part in floor cleaning schedules.

The word ' polish ' in the dictionary is defined as ' smoothness or glossiness produced by friction '. Developments in recent years, particularly in the field of water emulsion floor waxes have shown that friction, or buffing is not always necessary to produce a smooth, glossy finish. ' Polish ', associated as it is with buffing or manual effort, is thus in the context of floor maintenance, somewhat outdated.

A more modern, and appropriate, term is ' floor-wax ', which embraces all materials of this type used to maintain floors. Many have dry-bright qualities, that is they dry to a high gloss without any buffing or manual effort. Whilst the word ' polish ' has been in use for a very long time, modern technological developments are rapidly making it obsolescent.

There are two types of floor wax in use today, namely solvent-based wax and water emulsion floor wax. They will be considered separately.

SOLVENT-BASED WAX

History

In Anglo-Saxon times natural beeswax was known as ' weax ' from which the term ' wax ' was derived.

Solvent waxes were first used in the thirteenth century, when Italian floors made of marble, stone and terrazzo were waxed to improve their appearance. They were only employed on a very limited scale and it was not until the following century, when modern parquetry floors were introduced in France, that solvent wax became widely used.

At first beeswax was used, being melted into the floors with hot irons and then buffed to a high gloss by servants with rags wrapped around their feet. Later, to enable it to be dispersed more easily, the beeswax was dissolved in turpentine. This was the forerunner of paste waxes which are still in use. In 1797 Kosler, on an expedition to eastern Brazil, discovered carnauba wax on the leaves of palm trees. Manufacturers of solvent waxes found that by using carnauba wax in place of beeswax a much more durable material could be made, having a higher gloss.

With development of cheaper solvents manufactured from petroleum, the use of turpentine was eventually discontinued. Synthetic waxes have also largely replaced natural waxes, with the result that most modern solvent waxes consist almost entirely of synthetic materials.

Definitions

A ' solvent-based wax ' can be defined as a wax or blend of waxes dispersed in a solvent.

Waxes are amorphous masses, usually consisting of esters of long, straight chain fatty acids. The classification of waxes stems originally from basic beeswax, which has the formula: $C_{30}H_{61}O \cdot CO \cdot C_{15}H_{31}$.

This definition can be somewhat extended to include the large number of hydrocarbon waxes and various synthetic waxes now used in the industry. From a practical point of view, the definition of wax can be taken as ' a material which, either physically or chemically, resembles beeswax '. The term ' wax ', therefore refers more to the properties of a material than to its chemical composition.

A solvent can be defined as ' that component which is present in excess, or whose physical state is the same as that of the solution '. In explanation, it can be said that any liquid which will dissolve a solid is a solvent for that solid. In the case of solvent-based waxes, the waxes are dissolved by turpentine, white spirit or similar petroleum-based materials.

Floors

Solvents are generally harmful to a number of floors, including some of the more modern types of floor covering. For this reason solvent-based waxes can only be used on a limited variety of floors; they are recommended for use on wood, wood composition, cork, magnesite and linoleum. The presence of solvent, in addition to assisting the spreading of wax, helps, in some instances, to clean the floor as the wax is applied.

Solvent softens asphalt floors and has a detrimental effect and the waxes must not, therefore, be used on them. Similarly, it softens the asbestos filler used in the manufacture of thermoplastic tiles and PVC(vinyl) asbestos tiles, and can also dissolve the pigments in them, often causing the colours to run into each other.

Pure flexible PVC will resist solvent quite successfully and could be treated with a solvent wax, if required. There are, however, very few floor coverings which consist of pure PVC, the majority containing a relatively small proportion of some other material which is affected by solvent. It is always advisable therefore, to avoid using solvent-based waxes on these floors and to use a water emulsion floor wax instead.

It is seldom, if ever, required to obtain a glossy surface on concrete floors. If the floor is sealed, however, a good quality wax will provide a barrier against foot traffic and considerably extend the life of the seal. A solvent-based wax can be used for this purpose and will not harm concrete, although a water-based wax is preferred.

Quarry tiles are extremely hard and the presence of a solvent-based wax may produce slippery conditions, particularly if water is also liable to be present on the floor. For this reason it is inadvisable to treat the tiles with a solvent wax.

This applies also to terrazzo and marble floors where the wax not only tends to make the floors slippery but also to slightly darken light coloured floors.

Solvent-based waxes must not be applied to rubber floors because rubber is adversely affected by the solvent present in these materials. White Spirit and similar solvents cause rubber to become soft and sticky; colours may also dissolve and ' bleed ' onto adjoining areas, so ruining the appearance of the floor.

Requirements of a Solvent-based Wax

Since solvent-based waxes are used frequently and are not semi-permanent finishes, they must be formulated to satisfy different requirements from those of seals.

The original, primary requirement was for a floor wax that could be buffed to a high gloss. As technology advanced, so users and manufacturers became aware of a number of others, equally important, that must be fulfilled if solvent-based waxes are to play their part effectively as floor maintenance materials.

The main requirements of a floor wax are as follows. It should:

Prevent dirt from being ground into the floor surface.
Protect the floor from water, chemicals and stains due to spillage.
Provide a pleasing sheen or gloss.
Resist scuff marks.
Possess anti-slip properties.
Resist carbon black heel marks.
Be buffable after wear for gloss renewal.
Be durable.
Have good adhesion and resistance to flaking off.
Not alter the colour of the floor to which it is applied.
Not yellow with age.
Have good levelling and flow characteristics.
Have good overcoating properties.
Be easily removed, when required.
Be easy to apply.
Be quick drying.
Have indefinite shelf life.
Have a mild odour.
Have realistic cost.

In addition, paste waxes must also be firm in consistency and show no solvent separation on standing. Liquid waxes must be homogeneous in consistency and also show the minimum of solvent separation on standing and be easily returned to their original condition on shaking.

Fulfilment of these requirements depends very largely on the characteristics and qualities of the raw materials available. These will now be discussed.

Raw Materials

In recent years a very large number of synthetic raw materials, both waxes and solvents, have been developed. Some have largely replaced the natural materials originally used in the manufacture of solvent-based waxes, whilst others have yet to be proved.

The most important natural and synthetic waxes include the following.

Natural Waxes

Beeswax—Beeswax was used in considerable quantities in the early solvent-based waxes but has now become obsolescent in this field. It is produced by the worker bee and is a secretion formed in the bee's stomach from honey and flower pollen. Wax production is controlled by the bee keeper depending upon whether he requires wax or honey as the main product.

The wax is melted, filtered and sometimes bleached, so that the colour varies from light amber to a medium brown. Beeswax blends well with solvent and remains stable for long periods in this form. On its own, however, beeswax has poor buffing qualities and little durability.

Carnauba Wax—Carnauba wax is used in considerable quantities in the manufacture of solvent-based waxes, a small amount being sufficient to improve the gloss and wearing properties to a marked degree.

Carnauba wax is obtained from the leaves of the cera or carnauba palm, grown in Brazil. The amount produced annually is approximately 11,000–12,000 tons. The wide, long leaves are harvested twice a year. They are laid in rows for drying in the sun and, after about 5 days, when the fibrous matter has shrunk, the leaves are beaten with sticks to detach the wax, which is collected, melted and poured through filter cloths into moulds.

Two main grades are available, prime yellow and grey, a cheaper, less pure form.

Montan Wax—This wax is hard, has a high melting point, gives a very good gloss and has excellent solvent binding properties, in addition to good wearing qualities.

The wax is extracted with solvents from lignite or brown coal, mainly found in central Germany. It can be bleached or chemically treated to produce a range of useful waxes.

Other Waxes—Included among other natural waxes are Candelilla wax, a Mexican product derived from a weed of the same name. Ouricury wax is another Brazilian material, obtained from the leaves of a different type of palm tree from those producing carnauba wax. Palm wax, cotton wax and esparto wax are other natural, vegetable waxes. Although each has its uses, none is used as widely as carnauba wax.

Natural insect waxes include ghedda wax, obtained from India and East Asia, and shellac wax, obtained from a type of cochineal or lac insect, which has very good buffability and will produce a high gloss.

Synthetic Waxes

Paraffin Wax—Paraffin wax is used in considerable quantities in the manufacture of solvent-based waxes. Many different grades are available, all with different melting points and other properties. They have the advantage of being relatively inexpensive and are usually incorporated with better quality waxes. Used on their own they would yield a poor consistency material with low gloss qualities.

Paraffin wax is derived from petroleum. When petroleum crude oil is distilled a fraction is obtained consisting of lubricating oil and paraffin wax. The wax is separated from the blend by cooling, when it crystallizes out of solution. It is filtered and finally purified with sulphuric acid.

Polyethylene Wax—Polyethylene wax is used in small amounts in solvent-based waxes, although very large quantities are used in the manufacture of water emulsion floor waxes. It is included in solvent-based wax to improve the slip resistance qualities and to give better gloss and durability.

Other Synthetic Waxes—Among these are Ozokerite wax, also used in small quantities and Ceresin wax, derived from a bleached and purified ozokerite.

Solvents

A number of solvents are used in the manufacture of solvent-based waxes. The best known is, perhaps, turpentine which, although still used in shoe polishes, has largely been replaced in floor waxes by white spirit, sometimes known as turpentine substitute.

White spirit is derived from petroleum and is obtained by heating crude oil until it is partially vapourized. The vapour rises in a tall structure resembling a large tower. Trays are placed at equal intervals throughout the height of the structure. As the vapour rises, it cools. Various ingredients in the crude oil become liquid at different heights and are collected on the trays before being transferred to storage tanks.

At the bottom of the structure the heaviest fractions, consisting of fuel oil and lubricating oil, are collected. Other fractions, rising up the structure, include gas oil, kerosene or paraffin and naphtha, from which the white spirit is obtained. Light motor spirits are obtained from the top of the structure.

The bland odour of white spirit is one of its most useful properties. When properly manufactured it has a characteristic smell, less pungent and more pleasant to work with for prolonged periods than that of almost any other organic solvent. White spirit is perfectly colourless and can be classed as water-white.

The rate of evaporation is of considerable practical importance. It is essential that it should not only evaporate at a reasonable rate, but should evaporate completely at ordinary room temperatures. As white spirit is a mixture of hydrocarbons it has no definite boiling point. The range of temperature over which it distils should lie between 140°C and 230°C. The 'medium' grade generally used in solvent-based waxes has a range of between about 160°C and 200°C.

Certain cheaper solvents have an upper limit above this range. This higher limit implies a 'tail' of non-volatile hydrocarbons which retard the drying time of an application of solvent wax. Some cheaper solvent waxes have a tendency to dry slower and to retain a certain 'greasy' appearance, often still visible after the film has been buffed.

White spirit is sometimes used in conjunction with a faster evaporating solvent, used to accelerate the drying process. Whilst solvent-based waxes containing white spirit as the only solvent usually take about 20 to 30 min to dry, those including fast evaporating solvents can be made to dry in about 5 to 10 min.

White spirit and similar solvents are flammable, having a flash point of about 41°C (106°F). Whilst solvent-based waxes are perfectly safe under normal conditions of use, they should not be exposed to naked lights.

Additives

Other raw materials used in the manufacture of solvent-based waxes include perfumes, or 'masking agents', used to cover the smell of the evaporating solvent. Dyes are also included in very small amounts to colour the waxes, as many are colourless in appearance.

Silicone is generally included in furniture polishes, as it helps in the formation of a hard, glossy film and gives an element of slip to the surface. Due to the latter property, and also because floor waxes incorporating silicone are much harder to remove than those without, they are not generally included in solvent-based waxes.

Types of Solvent-based Wax

Solvent-based waxes are available in both paste and liquid, or free-flowing, forms. Each type will now be considered.

Paste Wax

The essential difference between paste and liquid wax is that the former contains a much higher proportion of wax. Paste wax generally consists of about 25–30% wax, the balance comprising solvent, colouring matter and perfume.

SOLVENT-BASED WAX

It is used to form a protective layer of wax on, generally, wood, cork, and linoleum floors, and also on magnesite floors.

The floor must be clean before the wax is applied. Ideally it should be new or sanded, so that there is no dirt present at all. Alternatively, all previous applications should be removed and the floor thoroughly cleaned and allowed to dry. If the area to be treated is small, paste wax should be applied to the floor with a cloth or applicator and spread with the brush of a floor polishing/scrubbing machine, so that the bristles force the wax into all the crevices.

If the area is large, a sprayer, with heating element, is recommended. If this is not available, the consistency of the floor wax can be reduced by the addition of an approved solvent. Great care must be taken when adding solvent because of the fire risk. The thinned paste wax should then be applied and spread with an electric polishing/scrubbing machine or applicator, forcing the wax into the open pores.

It is important that whichever method is chosen, the wax is applied evenly over the floor and rubbed in so that only a thin layer remains on the surface. This should then be buffed to raise a high sheen and to harden the surface and improve durability.

Once the open pores of the floor are filled with wax, further maintenance should be with a liquid wax. This is because the paste materials, with their high wax content, will not clean a floor, but merely entrap the dirt and enable it to be spread around. In time a build-up of thick, glutinous, wax takes place which, together with dirt present on the floor, darkens the surface and makes it both unsightly and unhygienic.

A thin application of paste wax will give better results than a thick one and will assist in the prevention of a build-up of wax.

When properly applied and buffed, paste waxes have relatively good slip resistance properties. This is, however, reduced if a build-up of wax is allowed to take place. They also resist carbon black heel marks well, because, when the shoe is applied to the floor, the wax gives a little and allows the heel to pass over it without carbon black becoming detached. For the same reason, however, paste waxes tend to show scuff marks, but these can be easily cleaned off and the gloss renewed by buffing.

Paste wax is easy to remove, when required. This should be done using a solvent-based detergent wax remover, specially formulated for this purpose, spread over the area to be treated with a brush or mopping equipment. The wax remover should be allowed to penetrate into the wax for a few minutes, after which the floor is scrubbed using a metal fibre or nylon web pad under an electric polishing/

scrubbing machine. Alternatively, if a machine is not available, the wax can be removed with mopping equipment.

After removal, the floor should be thoroughly rinsed with warm water and allowed to dry.

Liquid Wax

Liquid wax contains very much less wax than paste wax, the normal quantity present generally being about 8–12%. The remainder consists of solvent, colouring matter and perfume.

Liquid waxes are produced in two physically different forms. One is extremely free-flowing and the other is of a thixotropic nature, being 'floppy' in appearance. The thixotropic type becomes liquid on agitation or shaking, re-setting to its thick consistency on standing. It is always advisable to stir or shake liquid containers before use to ensure the wax is evenly distributed throughout the liquid.

Both types contain a similar amount of wax, the difference in appearance and consistency being due to the different properties of the waxes used. As both types behave similarly and give similar results they will be considered together.

Liquid wax is used in very large amounts on wood, cork and linoleum floors and, to a lesser extent, on magnesite floors. It is a very efficient cleaning agent. This is because the material is free flowing and the solvent component loosens the dirt on the floor as the wax is applied. The loosened dirt is retained in the applicator and a thin film of wax is deposited. When dry, the film should be buffed to harden the surface and to produce a gloss finish.

Liquid wax can be applied using mopping equipment or a lambswool bonnet, also by spray or dispensed from the tank of a polishing/scrubbing machine.

The material should be applied sparingly so that only a small quantity of wax is deposited on the floor, just sufficient to replace that removed by normal traffic. The edges of corridors, sides of rooms and other areas not subjected to traffic should be avoided, so that a build-up of wax does not take place.

When properly applied and buffed, liquid wax has good slip resistance properties which are, however, reduced if a build-up of wax is allowed to occur.

Liquid wax also resists carbon black heel marks well. Any scuff marks which may occur can be readily removed and the gloss restored to its original appearance by buffing, using a fine grade metal fibre or nylon web pad under an electric polishing/scrubbing machine.

Frequency of application will depend upon the type and volume of traffic. In general, one application each week should be sufficient, buffing as required in between. A common fault is to apply too much wax, resulting in a dirty floor. It is preferable to use less material and to buff frequently.

Provision should be allowed in every maintenance programme for periodic stripping to remove all old wax. This can be carried out using a solvent-based detergent wax remover, formulated for this purpose. The method is the same as that described for removal of paste wax, page 113.

Results obtained with both paste and liquid wax are very pleasing providing the floor is maintained in the correct way.

WATER-BASED WAX

History

With the introduction of rubber flooring and thermoplastic, or asphalt, tiles, problems of maintenance were introduced which could not be solved with solvent-based waxes. Such waxes contain white spirit, or other petroleum solvents, which soften these surfaces and cause the colours to bleed.

A considerable amount of work was carried out during the late 1920s in an attempt to emulsify hard waxes, thereby making it possible to carry the waxes in harmless water, rather than in solvents.

In 1929 the first water emulsion floor wax was marketed, having had its origin in the leather industry and being introduced as a floor wax by a leather chemist in St. Louis, U.S.A.

This started an industry which has grown very rapidly indeed. In the present day industrial field, water emulsion floor waxes account for approximately 80–85% of all such produced. The great increase in the use of these waxes has been due, in the main, to the continued successful development of a wide range of synthetic flooring materials, most of which are affected by solvents but remain unaffected by water.

The first water emulsion floor waxes consisted of carnauba wax together with a borax suspension of shellac in water. The shellac was included to impart hardness, toughness and increased gloss to the films. Later the shellac was replaced by other alkali soluble resins.

The years 1939–1946 saw no technological advances in this field. The industry was fully occupied with finding raw materials and with the pressure of other war work.

Immediately after the war, oxidized microcrystalline waxes were developed which were, in general, considerably softer than hard carnauba wax. They were blended with carnauba wax and shellac, or alkali-soluble resin, to produce a satisfactory water emulsion floor wax at a lower cost than could be obtained with carnauba wax alone.

Later, Fischer-Tropsch waxes were developed in Germany and, more recently, low molecular weight polyethylene was introduced.

In the early 1950s a new material emerged which surpassed all water emulsion waxes then on the market with respect to gloss. This was based almost entirely on shellac with about 20% wax. Unfortunately, however, products based on this formulation, and others like it, suffered from certain shortcomings. In particular, the emulsion tended to yellow on ageing and discolour pale floors. It also lacked durability and tended to 'water spot' badly.

Manufacturers experimented widely to find a material that could be included in an emulsion to overcome these defects and satisfy certain other requirements. It was found that a chemical group known as 'polymers' provided an answer to the problem and in the mid 1950s polystyrene was introduced to the industry.

More recent polymers include ethyl acrylate, methyl methacrylate and acrylic-polystyrene copolymers. Acid-sensitive polymers have also been used in limited quantities. Metal-containing polymers, conferring detergent resistant properties, are now widely used and research is continuing in this field.

Water-based waxes are used on a wide variety of floors, including thermoplastic tiles, PVC(vinyl) asbestos, flexible PVC, linoleum, asphalt, rubber, terrazzo and sealed wood, wood composition, cork and magnesite surfaces. They are recommended for use on almost all the new synthetic flooring materials currently being produced, including foam- and felt-backed PVC and similar materials.

Being aqueous, they are non-flammable.

Definitions

The term 'emulsion' frequently requires explanation when used in connection with floor maintenance materials.

An emulsion is a very fine suspension of one liquid in another liquid with which it is not normally miscible. For example, oil and water are not normally miscible and will separate if blended together. They can, however, be emulsified by the use of emulsifying agents which finely disperse and suspend one liquid in another.

Wax is emulsified in water by heating it, with emulsifiers and other materials, until it becomes liquid. It is then added to hot water, with stirring, if the 'wax to water' technique is being employed. In this state a true emulsion of one liquid in another is formed.

The wax then solidifies into very fine particles and remains dispersed in the water. By common use, although the wax is in a solid state, a suspension of wax in water is termed a 'water wax emulsion'.

Some explanation of the term 'wax' has already been given in the section dealing with solvent-based wax (page 107). Carnauba wax has now very largely been replaced in water emulsion floor wax formulations by other, cheaper, materials, many of synthetic origin. Polyethylene, for example, is referred to as a 'wax' although it is chemically a resin and not a natural wax material. Materials containing polyethylene and other, similar, raw materials, are referred to as either 'wax emulsions' or 'wax-free emulsions', depending upon whether polyethylene is termed a 'wax' or 'resin'. Both descriptions are correct, according to how the word 'wax' is interpreted. Considerable confusion has arisen in the past over descriptions. This play on words, however, is not important; it is the behaviour of the material under actual-use conditions that really matters.

An alkali-soluble resin is simply a resin that will dissolve in an alkaline solution. Shellac was used originally but a large number of synthetic materials are used for this purpose today.

The word 'polymer' also occurs frequently in connection with emulsion floor waxes. A polymer is a very large complex molecule formed by the reaction together of a great number of small molecules. Some compounds, for example styrene, are capable of combining with themselves. The result is a material which is very tough because the molecules are linked into giant groups. Polymers can be manufactured synthetically to form synthetic polymer resins. If the raw materials combine with other, like, materials they form simple polymers, for example, the polymerization of styrene to form polystyrene. If, however, another raw material is added it is possible to form copolymers, as in the polymerization of styrene with an acrylate to form a polystyrene-acrylate copolymer.

Early formulations consisted of two main ingredients only, a wax and an alkali-soluble resin or shellac. These materials are known as 'two component systems'.

Later, when polymers were included, the materials became known as 'three component systems'. The principal ingredients, wax, alkali-soluble resin and polymer, can be blended in almost any proportions to give emulsion waxes with a wide variety of properties.

Floors

Because emulsion floor waxes are water based and contain no solvent, they can be used on a much wider variety of floors than solvent-based waxes. They are used very successfully on thermoplastic tiles, PVC(vinyl) asbestos, flexible PVC, rubber, asphalt, linoleum, terrazzo and marble floors. They are also used on magnesite floors; although magnesite floors laid in this country commonly have a wood filler, the amount of water coming into contact with the floor is so small as to have no effect, particularly if it has been sealed.

One of the main reasons for the rapid rise in demand for water emulsion floor waxes is their suitability for use on such a wide range of flooring surfaces. This can be a great advantage, particularly in buildings with different types of floors. Many such buildings are very difficult to maintain, because each type of floor requires a different maintenance schedule. Rather than use two different types of wax on limited areas, it is often preferable to use one throughout. Because of this, the use of water emulsion floor waxes has grown very rapidly, often at the expense of solvent-based wax.

Linoleum is a flooring that can be maintained equally well with either a water emulsion floor wax or a solvent-based wax. If the building contains other floors, for example thermoplastic tiles, that must be maintained with a water-based material only, it is preferable to standardize on a water emulsion wax throughout.

Another factor to be taken into account when considering a water- or solvent-based wax for linoleum is resistance to slip. In general, water-based materials have a greater resistance to slip than solvent-based ones. If slip is important, a water emulsion floor wax is preferred.

Asphalt, particularly, and magnesite, occasionally, lose colour and fade on ageing. A coloured water emulsion wax will remedy this and maintain the floor in its original, attractive condition.

It is seldom required to maintain concrete floors with a wax, even when they are sealed. A good quality emulsion floor wax will, however, provide protection against traffic and extend the life of the seal.

Water-based emulsion floor waxes are not, in general, recommended for use on quarry tiled floors. Quarry tiles are extremely hard and although an emulsion floor wax can be successfully applied there is little advantage in doing so. Cleaning with an approved detergent is generally quite sufficient to maintain them in a satisfactory condition.

Water-based emulsion is not satisfactory for application to wood, wood composition and cork floors, to which water is detrimental. Constant use of water will remove the natural colour, cause the surface fibres to splinter and detach from the floor and carry dirt into the surface.

There has been a considerable difference of opinion regarding the use of a water emulsion wax on a sealed wood, wood composition or cork floor. A good seal, correctly applied, should be impervious to water. Consequently if the seal is intact, water will not come into contact with the floor itself, but will remain on the surface of the seal. If, however, the seal becomes worn, perhaps at the entrances to rooms or in main traffic lanes, water may penetrate into the floor carrying dirt with it. The result is a blackening of the floor in the worn areas where seal is no longer effective. This looks unsightly and the affected area must be sanded and re-sealed at the earliest possible opportunity. If a solvent-based wax is used on these floors and the seal wears through, solvent wax comes into contact with the floor itself. This is perfectly satisfactory as a solvent-based wax is recommended for use on these floors.

A water emulsion floor wax can, therefore, be used with absolute confidence on a well sealed floor. The seal must, however, be inspected from time to time and a further coat applied in the main traffic areas before it becomes worn.

If there is any doubt about the condition of the seal, a solvent-based wax is preferred.

Requirements of a Water Emulsion Floor Wax

Water emulsion floor waxes are applied to a wide variety of floors of differing colours. The modern trend is towards light coloured floors, in many cases pure white. This movement towards paler shades has had a considerable effect upon the relative importance of certain requirements. Light coloured, non-yellowing properties are now an essential of any water emulsion floor wax.

As waxes form the surface in direct contact with traffic, it is essential that any damage to the floor wax can be easily repaired. Unlike seals, water emulsion floor waxes must be stripped periodically from the floor. One of the most important requirements, therefore, is ease of removal.

The main requirements of a floor wax have already been listed on page 109.

There are, in addition, other requirements, many of a technical nature, and a single product cannot fulfil them all. For this reason a number of types of water emulsion floor wax have been developed, each with different characteristics. In this way, products are available to satisfy each different set of circumstances. By varying the formulations certain basic requirements can be accentuated, at the same time maintaining other requirements at a reasonable level of performance. For example, some products are formulated to have excellent anti-slip properties, whilst other desirable properties are maintained. Other products are formulated to be extremely easy to remove, so that this can be done without the use of machines and using female labour.

To a few organizations, low cost is of the utmost importance. In general, the lower the cost, the lower is the standard of performance that can be expected from a floor wax. Extremely cheap products, therefore, compromise some of the desirable properties to achieve low cost and are, in the long term, often less economic than more expensive products.

The extent to which requirements can be met depends to a large degree on the properties and characteristics of raw materials used in water emulsion floor waxes. These are discussed in the next section.

Raw Materials

A considerable amount of development work is being undertaken both by floor wax manufacturers and by raw material suppliers to improve existing materials and develop new ones. The floor wax industry has grown very rapidly; the technology is still in its infancy, but great strides forward have taken place in recent years. Development in some fields of activity, particularly with regard to polymers, is proceeding at such a pace that in some instances newly developed materials are obsolescent even as they become available on the market.

There are about fifteen different raw materials in some of the more sophisticated water emulsion floor waxes in use today. The main ones belong, however, to four groups, namely:

(1) Waxes.
(2) Alkali-soluble resins.
(3) Polymer resins.
(4) Additives.

Each of these is dealt with in some detail, below.

(1) *Waxes*

The wax component is of considerable importance in any water emulsion floor wax formulation; the type and quantity deserves and receives very careful consideration. The wax contributes to the durability and provides flexibility and buffability. It is the component of the floor wax that moves when subjected to traffic. If the film is extremely hard and unyielding, the sole or heel of a shoe will be brought to a halt with a jerk. This will cause a fraction of the heel, particularly, to disintegrate, resulting in a carbon black heel mark remaining on the floor.

If, on the other hand, the wax is extremely soft, it will give on impact, resulting in a floor free from carbon black heel marks, but badly scuffed and showing little, if any gloss. A wax must, therefore, be chosen either with intermediate properties or in combination with another wax with different, complimentary, properties.

(1.1) *Carnauba Wax*—This contributes high gloss, toughness and durability to the film of floor wax.

The origin of Carnauba wax has been described earlier (page 110). It is available in two main grades, prime yellow and grey; the grey material is a cheaper, less pure form and has a tendency to produce dark emulsions, which can discolour the floor if a number of coats are applied.

Prime yellow material is preferred because it is very much lighter in colour. It is, however, more expensive than the grey material and is usually blended with other, synthetic waxes to reduce the overall cost.

(1.2) *Montan Wax*—Montan wax can be used with advantage in water emulsion floor waxes; it is hard, gives a good gloss and has good buffable properties. The wax is easily emulsified and can be incorporated in the manufacture of non-ionic and anionic dry-bright formulations.

Montan wax is obtained from lignite or brown coal, mainly found in central Germany. It is generally bleached or chemically treated to produce a range of useful raw materials, with varying characteristics.

(1.3) *Microcrystalline Wax*—The advent of microcrystalline wax in water emulsion floor wax formulations was largely brought about by the limited supply of good quality carnauba wax. Microcrystalline waxes are high melting point materials derived from petroleum oil.

There are a very great number of different qualities in use today. They are not suitable for use on their own as they tend to be very hard and are not as durable as some of the other waxes. Some oxidized microcrystalline waxes are used as a partial replacement for more expensive waxes to cheapen the raw material costs and are particularly suitable for use in water emulsion floor waxes as they are easily emulsified. Oxidized microcrystalline waxes are light in colour, very hard and give a good gloss to the floor wax film.

(1.4) *Polyethylene Wax*—Polyethylene is synthesized by the polymerization of ethylene together with a suitable catalyst. A wide range of polyethylene materials is produced. Although they are strictly resins in the chemical sense, it is convenient to include them among the synthetic waxes.

Low molecular weight polyethylenes are used extensively in water-based waxes. One of their most important properties is that they impart outstanding slip resistance qualities to the floor wax. They are also hard and give a better gloss than many other waxes. Polyethylene is also practically colourless, an important property where light coloured floors are concerned.

It has been found from experience that the proportion of polyethylene wax present should not be greater than about 25%. If this percentage is exceeded some floor wax films tend to become soft and tacky. An excessive amount of polyethylene could also lead to 'powdering' difficulties. The demand for polyethylene has grown so rapidly in recent years that it is now probably the most widely used synthetic wax.

(1.5) *Other Waxes*—Other waxes include Fischer-Tropsch waxes, produced in a wide variety of hardnessess and melting points. Many modified montan waxes are also available, having greatly differing melting points and performance characteristics.

(2) *Alkali-Soluble Resins*

The alkali-soluble resin component is usually present at a level of about 5–15% of the total water emulsion floor wax formulation and is included to aid levelling, or flow, and to contribute towards gloss and durability. Because the materials are soluble in alkaline solutions they also aid removability, perhaps their most important function.

Formulations from which alkali-soluble resin is omitted are generally extremely difficult to remove, as alkaline detergents have no point of entry into the floor wax film. If, however, it is present in excess, the film may have a tendency to be brittle and to scratch rather easily. By including about 5–15% alkali-soluble resin optimum results are obtained.

The first type of alkali-soluble resin solution consisted of shellac dissolved in ammonia and this was blended with the emulsified wax. Although levelling and gloss were improved, the shellac tended to water spot rather badly and, as the film aged on the floor, polymerize so that it became dark in colour and difficult to remove. A considerable amount of work has been carried out, both to improve shellac to its present, highly satisfactory, level and to develop new, synthetic, materials.

It would not serve any useful purpose to detail the physical properties, advantages and qualities of each type of synthetic alkali resin in use today. They include rosin maleic adducts, terpene phenolics, styrene maleic resins, polyesters, resin-modified polyesters, modified pentaerythritol resin esters, maleic resin esters and many others.

When formulating water emulsion floor waxes the alkali-soluble resin is selected not only for the physical properties mentioned above, but also for its compatibility with the wax component and particularly with the polymer. The behaviour and performance of a polymer can be greatly influenced by the alkali-soluble resin blended with it. If high gloss and good film forming properties are to be obtained, discrimination in selection of the alkali-soluble resin is of the utmost importance.

(3) *Polymer Resins*

The addition of a polymer resin to a blend of emulsified wax and alkali-soluble resin has been one of the most important developments in the floor wax industry in recent times. This addition gives to the wax considerably increased durability, particularly under heavy traffic conditions. In addition, by selecting the appropriate polymer, the gloss, water resistance, toughness, flexibility and removability can all be increased or varied according to particular requirements.

The earlier types of water emulsion floor wax were primarily designed for use on asphalt, rubber and thermoplastic surfaces. The inclusion of a water-based polymer resin improved the products used on these floors and extended their recommended use to linoleum, PVC(vinyl) asbestos, flexible PVC, terrazzo and sealed wood, cork and magnesite floors, in addition to those floor surfaces mentioned above.

The type of polymer present in water emulsion floor wax is important, as the physical properties and behaviour of the wax depend largely on the type and amount of polymer present. That this is widely recognized is evidenced by the very considerable amount of development work carried out in recent years on improving existing polymers and developing new ones.

Whilst the type of wax and alkali-soluble resin to some degree influence the performance of a water emulsion floor wax, the character and performance of the polymer is by far the dominant factor. This is because polymers vary widely in their physical properties and comprise usually by far the largest component. In dry-bright materials for example, the amount of polymer present is normally double the quantity of wax and alkali-soluble resin put together.

The main types of polymer available for use in water emulsion floor waxes, together with their properties, will now be discussed.

(3.1) *Polystyrene Polymer*—Polystyrene is manufactured by polymerizing the styrene monomer. Ultra-fine emulsions of less than 0·05 μ particle size can be produced. This is important from the gloss aspect, as the finer particle size materials tend to have a higher gloss.

Water emulsion floor waxes with polystyrene as the main polymer component tend to form hard films with high initial gloss. They are, therefore, widely used in the dry-bright type of formulation. They are durable and will withstand successfully heavy traffic conditions. They are freeze–thaw stable and are also stable at elevated temperatures.

The earlier polystyrenes tended to be amber in colour, rather than white and to degrade in ultra-violet light with the result that a number of coats discoloured the floor.

The manufacturing technique has, however, improved very considerably in recent times so that modern polystyrene materials are almost water-white in colour and do not discolour on ageing.

Floor waxes with a high percentage of polystyrene, therefore, are characterized by high initial gloss, hard films and good durability. They are not, however, as easy to remove and do not retain their light colour to quite the same extent as some floor waxes based on other polymers.

(3.2) *Acrylate Polymer*—Acrylate polymers are manufactured by the polymerization of a number of materials including methyl acrylate, ethyl acrylate, butyl acrylate and methyl methacrylate. The particle size can also be produced in the order of 0·05 μ.

Water emulsion floor waxes with an acrylate as the polymer component have different characteristics from those based on polystyrene. They have a tendency to remain on the surface of the floor and form good films without modification. One of the results is that the films appear to have a greater depth of gloss than is the case with polystyrene materials. The refractive index of acrylate is, however, less than that of polystyrene materials, so that they tend to produce less gloss at equal solids content, although the depth of gloss, as mentioned above, is greater.

Acrylates are pure white in colour and do not yellow on ageing, a considerable advantage when used on white or light coloured surfaces. They are also more flexible and are softer and tougher than polystyrene materials. They have no tendency to powder and, perhaps one of their greatest assets, they have very good resistance to carbon black heel marks. Other benefits include excellent resistance to water spotting and, when formulated correctly, ease of removal.

Water emulsion floor waxes based on acrylate polymer tend to be less freeze–thaw stable than the equivalent based on a polystyrene. For this reason, water emulsion floor waxes must be protected from frost.

(3.3) *Styrene-acrylate Copolymer*—When two different monomers are polymerized together the resulting material is called a copolymer. Copolymers are, perhaps, the widest used of this type of material and of these styrene-acrylates are by far the most important.

A satisfactory polymer can be obtained by blending three or four parts polystyrene to one part polyacrylate. Copolymers formed in this way combine the good properties of both polystyrene and polyacrylate materials in a single product. This is often more effective than blending the two, separate, polymers.

Water emulsion floor waxes with styrene-acrylate copolymer as the main constituent tend to have high surface gloss, very good durability and are extremely light in colour. They do not yellow on ageing and are easy to remove. Resistance to carbon black heel marks is of a high order and water spot resistance is also good.

Whilst copolymers have some freeze–thaw stability it is always advisable to protect them from frost.

(3.4) *Metal-complexed Polymer*—The formulation of complex polymers including a metal, often zinc or zirconium, is a comparatively recent development in the floor maintenance industry. Polymers including a metal have, however, been known for over one hundred years and have been used extensively in the leather industry. In the case of water emulsion waxes the aim is to give a greater degree of control over removability. When a conventional water emulsion floor wax is scrubbed with an alkaline solution, the alkali-soluble resin component breaks down enabling the whole film to be removed.

Metal-complexed polymers are designed for use without this resin. The reactive group consists of the metal complex in the polymer itself, so that films are not removed when scrubbed with an ordinary alkaline detergent. If, however, a detergent with a high pH value is used, or an ammonia solution, the metal complex breaks down enabling the film to be easily removed.

Water emulsion floor waxes incorporating these polymers have good gloss and good resistance to carbon black heel marks and to water. They are also very durable and light in colour and do not yellow on ageing. Their main characteristic is a resistance to neutral and alkaline detergents, so that the floor can be repeatedly cleaned with detergent without affecting the durability. They are, in general, not freeze–thaw stable and must be protected from frost.

(3.5) *Other Polymers*—Other polymers used in the formulation of water emulsion floor waxes include polyvinyl acetate, widely employed in the paint industry, which is sometimes used in small amounts to improve the toughness of fully buffable materials.

Acid-sensitive polymers have been used to improve detergent resistance properties, but the advisability of using an acid stripper on a floor is very questionable and these polymers have now largely been replaced by metal-complexed polymers.

(4) *Additives*

Although additives are only used in small amounts, selection of the correct materials is vital if the performance of a water emulsion wax is to be of the highest quality.

Blending a polymer wax and alkali-soluble resin will not, necessarily, produce the desired results. Additives, included at various stages in the manufacture of a water emulsion floor wax, influence film-forming properties, levelling and flow, powdering characteristics, removability, gloss and recoatability.

Additives can be divided into three main groups, (1) plasticizers, (2) wetting agents and (3) coalescing agents.

(4.1) *Plasticizers*—The selection of a plasticizer depends on its ability to lower the temperature of a polymer to enable it to form a film, hard polymers requiring more plasticizer than soft ones. In general, acrylate polymers require no plasticizer whereas polystyrene polymers must be plasticized, if not they have a tendency to become brittle on ageing. The presence of a plasticizer will prevent embrittlement and promote flexibility of the floor wax film. Plasticizers also influence the depth of gloss of a film, recoatability and resistance to powdering.

Satisfactory plasticizers are dibutyl phthalate and tributoxyethyl phosphate, used either separately or blended together. Others include tricresyl phosphate and benzylbutyl phthalate.

(4.2) *Wetting Agents*—Wetting agents are included to assist emulsification of wax and to promote stability in the finished material. They can vary from ammonia, in the form of triethanolamine, to more complex materials such as 2-amino-2-methyl-1-propanol. Materials that have achieved prominence in recent years are the fluorochemical surfactants. These are very effective at promoting levelling and flow at extremely low concentrations. As little as 50 to 100 p.p.m. are required to make quite a significant improvement in the appearance of the film.

Wetting agents also promote even spreading of the film so that it levels or flows evenly. Good flow contributes to the gloss of the film and to its adhesion to the floor.

(4.3) *Coalescing Agents*—These are added to water emulsion floor waxes to improve the surface-film forming properties. They act by softening hard particles of polymer, so promoting an even flow over the surface.

Coalescing agents are only used in very small amounts. The most commonly used materials are diethylene glycol, carbitol and 2-pyrrolidone.

The modern polymer emulsion floor wax is a blend of a considerable number of different raw materials, each selected to ensure that the finished product is scientifically exact. As many as twenty different materials can be incorporated in a single formulation. A correct blend is essential to ensure that the inherent properties of the floor wax are properly balanced and that the desirable characteristics are fully developed.

Types of Water Emulsion Floor Wax

It has been noted earlier that the requirements of a water emulsion floor wax are many and varied and that a single product cannot fulfil them all. Each user, furthermore, may have different specific requirements, one desiring a material with slip resistance as the most important characteristic, while another may consider non-yellowing as most important.

Different materials are, therefore, manufactured to satisfy the varying needs. In general, materials are formulated to meet specific requirements, so that certain qualities are highlighted whilst others remain at an acceptable level. In addition, technology is advancing all the time, so that the overall standard is improving slowly, yet continuously. An example of this is the present degree of resistance to yellowing on ageing of water emulsion floor waxes.

The first water emulsion floor waxes showed marked yellowing, even after only a few weeks. Today, this has been very largely overcome and rarely is yellowing a problem, even after many months on a floor.

Of a very large number of water emulsion floor waxes available, all fall into one of three main types: (1) fully buffable, (2) semi-buffable, and (3) dry-bright.

These types are relative to the performance of a material, rather than the formulation. They can, however, be very closely related to the respective formulations in that fully buffable materials are wax rich and the dry-bright materials polymer rich, with the semi-buffable floor waxes in between.

Approximate formulations are as follows:

	Fully Buffable (Wax rich)	Semi-buffable —	Dry-bright (Polymer rich)
Polymer %	20–40	45–60	50–70
Wax %	45–60	25–40	5–15
Alkali-soluble resin %	5–15	5–15	5–15

The earlier fully buffable materials were based on wax and alkali-soluble resin alone, but almost all now include a small percentage of polymer to improve durability.

(1) *Fully Buffable*

Fully buffable emulsion floor waxes normally dry to a low sheen which can be increased to a high gloss by buffing. The film is generally rather soft and buffing is essential not only to raise the gloss but also to harden the film and thereby increase durability.

Fully buffable materials generally have good resistance to carbon black heel marks and have fair scuff resistance but resistance to water, stains and dirt will generally be less than semi-buffable or dry-bright types.

The biggest advantage of this type of material is the ease of gloss renewal. When the film shows signs of traffic and scuff marks are evident, buffing the floor quickly restores the original high gloss. This can be repeated, either daily or weekly, as required, until it is evident that there is insufficient wax on the floor and a further coat is needed.

(2) *Semi-buffable*

Water emulsion floor waxes of this type, with more polymer and less wax, dry to a subdued gloss which can be increased, if required, by buffing. In many instances the initial gloss is considered to be perfectly satisfactory and the surface is left unbuffed. Buffing, however, hardens the film and increases durability and it is recommended that initial coats, particularly, should be buffed.

Semi-buffable materials have good resistance to carbon black heel marks. They are more scuff resistant than the fully buffable materials and have improved resistance to water, stains and dirt. Because of the high proportion of polymer present they are generally more durable than the fully buffable materials.

Scuff marks can be removed by buffing and the film repeatedly buffed to restore the original appearance.

(3) *Dry-bright*

Dry-bright emulsion floor waxes dry on application to a high initial gloss, without buffing. The true dry-brights, containing minimum wax, will show very little, if any, increase in gloss on buffing. They start, therefore, looking their best and gradually deteriorate under traffic.

Dry-bright materials are very durable and have excellent resistance to water, stains and dirt. They have, in general, better anti-slip properties than the more buffable materials, because the polymer will not give to the same extent as a wax surface when subjected to foot traffic. For this same reason, resistance to carbon black heel marks is not as good as fully buffable or semi-buffable materials and the black marks are rather more difficult to remove, as the film is harder.

Some of the first polymers produced tended to yellow badly on ageing, with the result that many polymer rich materials were discontinued. Modern polymers, however, are very much improved, so that an incidence of yellowing is extremely rare. Acrylic polymers, in particular, are water-white in colour and remain so even after long periods.

It is true that dry-bright emulsion floor waxes are generally harder to remove than either fully buffable or semi-buffable materials, because the films are harder for the alkaline detergent to penetrate. Modern formulations compensate for this by increasing the proportion of alkali-soluble resin accordingly, so facilitating break down of the film by alkaline detergents.

In general, greater physical agitation, either manual or using machines, is required to remove dry-bright floor waxes. The detergent should be allowed to act on the floor wax for several minutes before scrubbing commences, so that the chemicals in the detergent can effectively break down the wax film.

It must be appreciated that whilst the above are the three main types of water-based wax available today, the majority of products lie between these types and combine the properties of more than one type.

For example, many products combine dry-bright characteristics with buffable properties, which, once applied, dry to a good gloss and are re-buffable, when required. In the industrial field, particularly, some degree of buffability is necessary, so that the surface can be easily renewed without the necessity of applying a further coat.

Speciality Products

A number of speciality products have been developed in recent years to meet specific requirements. Whilst they can also be included among the fully buffable, semi-buffable or dry-bright types, they have special characteristics and will be considered separately.

The most important ones are (1) high-solids emulsion floor waxes, (2) wash-and-wax emulsion floor waxes, (3) acid-sensitive emulsion floor waxes and (4) detergent-resistant emulsion floor waxes.

(1) *High-solids Emulsion Floor Waxes*—The solid content of conventional emulsion floor waxes is generally in the region of 12–15%. This means, therefore, that when a water emulsion floor wax is applied only a proportion remains on the floor and the remainder, consisting essentially of water, evaporates from the surface. It has been found from long experience that a solid content of about 12–15% gives optimum results for conventional materials. The high proportion of water present is essential to enable the material to flow evenly over the floor surface and to provide a thin, even film of floor wax, so essential for good gloss and durability. With regard to conventional water emulsion floor waxes, two thin coats give infinitely better results than one thick one.

There are, however, occasions when a thick film is required rather than a thin one, for example, heavily trafficked areas only available for cleaning and wax application at infrequent intervals. Busy corridors in large factories, where traffic rapidly wears away conventional emulsion floor wax, are typical of those areas where a thicker, more durable, film is required.

High-solids emulsion floor waxes have been formulated to satisfy conditions where exceptional durability is required. As the name implies, they contain a very high solid content, approximately twice that of conventional emulsion floor waxes and in the region of 25–30%.

Careful formulation is essential to ensure that good flow and levelling are obtained. High-solids emulsions are not simply conventional materials with less water, but are designed specifically for the purpose of providing a thick film, approximately equivalent to two thin films, at a single application.

Because of the high solid content, materials of this type have excellent dry-bright properties with only one coat. If necessary, a second coat can be applied as soon as the first one has dried hard, usually about 1 h. The drying time of these emulsions is generally longer than that of conventional emulsion floor waxes, which dry in about 20 min, because of the extra thickness of the film. Emulsion floor waxes dry by evaporation of water, and this takes longer from a thick film than from a thin film.

High-solids emulsions have excellent slip resistance properties due to the thickness of the film. They can also be buffed repeatedly to restore the original gloss.

When using these emulsions, care must be taken to ensure that the material is not over-applied, particularly in the areas not subject to traffic. Due to the high solids content a build-up can quickly occur if areas, for example the edges of corridors, are over-waxed. Whilst a build-up can be removed with an alkaline detergent and, if necessary, abrasive nylon mesh discs, prevention is always better than cure.

(2) *Wash-and-Wax Emulsion Floor Waxes*—Wash-and-wax, also known as 'clean-and-shine', emulsions are a comparatively recent development in the field of emulsion floor waxes.

They consist of a combination of wax and detergent balanced to permit washing and waxing the floor in a single operation. The normal method of cleaning and waxing a floor involves cleaning with a detergent solution, rinsing and application of floor wax, followed by buffing if a buffable material is used. If a wash-and-wax material is used this can be reduced to a single operation. The detergent component loosens the dirt, which is retained in the applicator or pad, if a scrubbing/polishing machine is being used. The material is levelled and spread evenly over the floor and allowed to dry. The dry-bright component then takes effect, with the result that the floor is cleaned and waxed in one operation. Whilst this method has some limitations, which will be discussed in this section, it has much to commend it. Where wash-and-wax materials have been used successfully, very substantial savings in both labour and materials have been effected.

Wash-and-wax emulsions are not designed to remove excessive soilage or to rapidly convert a neglected floor to a high standard. They are designed to maintain a good finish under light to normal traffic conditions. Because they are comparatively recent developments they are by no means standardized, and a very wide range of materials can be found described as wash-and-wax emulsions.

If, in such a product, excess detergent is present, it can cause extreme bubbling or foaming on application and can also produce a film with low gloss, poor levelling and flow, a patchy finish and a soft, non-durable surface. If not enough detergent is present, the emulsion will not remove the dirt and the floor quickly becomes dirty in appearance. Discrimination in selection is, therefore, extremely important if the best possible results are to be obtained.

The emulsion component, in general consisting essentially of a polymer and wax, should have good dry-bright properties and should be buffable for gloss renewal. In the best quality emulsions the polymer is an acrylic material characterized by lightness of colour and non-yellowing properties.

Some materials, described as 'wash-and-wax' emulsions, consist basically of detergent with very little, if any, polymer. A little wax may be included, which remains on the floor after cleaning and must be buffed to produce a satisfactory finish. They have no dry-bright qualities and can only be used in areas where ample buffing equipment is readily available.

Because of their composition, wash-and-wax emulsions are sensitive to water and have a tendency to water spot. For this reason they should not be used in entrance halls and other areas where they may be subjected to water carried into a building on the soles of shoes. Due also to this sensitivity, however, each application softens the previous one to some extent, so that approximately half the earlier application is removed each time a new coat is applied. Not only, therefore, is dirt removed, but by removing part of the earlier coat a build-up of emulsion is effectively prevented from taking place. Maintenance can be carried out over an extended period of time without the need for stripping all old emulsion from the floor. This, again, reduces labour and material costs.

Absence of build-up and the use of acrylic polymer resins ensure that the lightest coloured floors remain light coloured, clean and attractive. Wash-and-wax emulsions are used with every confidence on even white tiled floors.

The emulsions are easy to remove when required. The lack of build-up and the sensitivity of the film to water and detergents ensure that when periodic stripping is eventually necessary it can be easily carried out.

Wash-and-wax emulsions generally have excellent resistance to carbon black heel marks and good resistance to scuff marks. They are often used as maintenance materials on top of water-based seals, particularly on very porous floors.

They are used widely in factory offices, away from main entrances, or on the second or higher floors of multi-storey office blocks and are employed to their greatest advantage under normal-to-light traffic conditions. When used properly considerable savings in both labour and materials can be effected.

(3) *Acid-sensitive Emulsion Floor Waxes*—A considerable amount of research has been carried out over a number of years to combine resistance to detergents with ease of removal. Any material with increased resistance to detergents and water is also very much more difficult to remove by conventional methods. Conversely, a material that can be very easily removed generally has poor resistance to both detergents and water.

Materials based on conventional polymers are, therefore, formulated to obtain an optimum balance between these two desirable properties.

In 1961 a special polymer was introduced which, it was claimed, had excellent resistance to detergents but which could be easily removed when required. Materials formulated using this polymer could thus be repeatedly scrubbed and would have greatly increased durability. The special polymer contained chemical groups known as 'amines', which break down when attacked with mild acids to effect removal. Emulsion floor waxes based on this type of polymer are called 'acid sensitive' or 'acid removable' emulsions. The chemical structure of this type of material is shown below.

$$R_1-N\genfrac{}{}{0pt}{}{R_2}{R_3} + HX \rightleftharpoons R_1-N\genfrac{}{}{0pt}{}{R_2}{R_3} .H + X^-$$

Insoluble polymer   Acid   Soluble polymer salt

Acid-sensitive materials have achieved a limited degree of success in the industrial field, although they have been largely superseded by metal-containing detergent-resistant materials, described in the next section.

A high degree of resistance to detergents is achieved using acid-sensitive polymers and durability is also much improved over conventional materials.

Acid-sensitive emulsions, have, however, four disadvantages and these must also be taken into consideration.

First, the polymers are more expensive to produce than conventional polymers, with the result that the emulsions are considerably higher in cost.

Secondly, a special acid stripper is necessary to remove the emulsion from the floor when required. Users must, therefore, stock another material, with the attendant difficulties of stock control, storage space and additional capital tied up in stock.

Thirdly, since the finish is acid sensitive, it can be damaged by such items as milk, fruit juices and other acid substances. This is, perhaps, a minor factor but under certain circumstances it could prove important.

Fourthly, and perhaps most important, the acid stripper, although mild, sometimes damages the floor itself. Most floors are harmed by acid and the use of acid as a normal maintenance material is not to be recommended. In some cases metal furniture has been found to be corroded by contact with acid.

Whilst acid-sensitive emulsions have been, and are being, used very successfully in some areas, it must be recognized that they have limitations.

(4) *Detergent-resistant Emulsion Floor Waxes*—It has been mentioned in connection with acid-sensitive emulsion floor waxes, that a great deal of research has been carried out to improve the resistance of polymers to detergent, whilst at the same time enabling the materials to be easily removed when required.

Metal-containing polymers, recently introduced to both the domestic and industrial markets, have proved extremely successful in the formulation of detergent-resistant emulsions which can, at the same time be easily removed when required.

Whilst metal cross-linked polymers are completely new concepts in the field of floor maintenance, metals have been used to cross-link polymers for over one hundred years. A typical application is in the tanning of leather, using chromium as the metal.

Emulsions based on metal-containing polymers have true detergent-resistant properties and can be easily removed when required. This is because they are sensitive to alkaline detergents of a specific pH value, or to ammonia. For this reason, some detergent-resistant emulsions are called 'ammonia sensitive' materials.

The chemical structure of metal-containing polymers can generally be represented as follows:

$$R_1—COO—Me—OOC—R_2$$
Metal containing polymer

WATER-BASED WAX

The metal component is generally zinc or zirconium. Most polymers are supplied with the metal present in the polymer, whilst others are prepared without metal, this being added during the manufacture of the emulsion. In either event, careful formulation is essential for success.

The behaviour of a metal-containing polymer emulsion in the presence of a conventional detergent solution is shown below. No reaction takes place as the emulsion is not affected.

$$\underset{\text{Metal-containing polymer}}{R_1\text{—COO—Me—OOC—}R_2} + \underset{\substack{\text{Alkaline} \\ \text{detergent}}}{M\cdot OH} \longrightarrow \text{No reaction}$$

If, however, the alkaline detergent is at a specific pH value, generally about 11·5, or contains ammonia, the polymer is broken down and the film can be easily removed. This is shown as follows:

$$\underset{\substack{\text{Metal-containing} \\ \text{polymer}}}{R_1\text{—COO—Me—OOC—}R_2} + \underset{\substack{\text{Alkaline} \\ \text{detergent} \\ \text{with ammonia}}}{M\cdot OH + NH_4OH} \rightarrow$$

$$\begin{cases} R_1\text{—CO}\bar{\text{O}}\ M\cdot NH_4{}^+ \\ R_2\text{—CO}\bar{\text{O}}\ M\cdot NH_4{}^+ \end{cases} + \underset{\substack{\text{metal} \\ \text{ammonium} \\ \text{complex}}}{} + \underset{\text{Water}}{H_2O}$$

Water-sensitive polymer salts

In addition to their detergent-resistant properties, emulsions based on metal-containing polymers have a number of other advantages. Their alkaline resistance enables rapid recoating without danger of harming the underlying film. The best materials are almost water-white in colour. Films are produced which do not yellow or discolour in any way on ageing. They can, therefore, be used with absolute confidence on even the lightest coloured floors. Ultra-violet light does not affect removability to any greater extent than conventional emulsion floor waxes.

Detergent-resistant emulsions are produced in both dry-bright and buffable qualities. The majority of industrial materials combine these qualities so that, whilst initial gloss is slightly lower than that given by a conventional dry-bright emulsion, gloss can be increased by buffing. Also, when scuff marks begin to show the film can be readily renewed to restore the original gloss. Detergent-resistant emulsions also show very good resistance to carbon black heel marks.

Perhaps the greatest single advantage they have over conventional emulsion floor waxes is improved durability. The life of the detergent-resistant emulsion film is much greater due to its resistance to scrubbing with water and detergents. This has been proved in practice and is now an established fact. Because of its increased durability, there is considerable saving in labour and materials with a consequent reduction in maintenance costs.

Discrimination in selection is, however, essential, as detergent-resistant emulsions may cause problems. Whilst polymers correctly formulated are cross-linked to a predetermined degree, some continue to cross-link even after application. The result is that the film of floor wax becomes harder and harder and even after a relatively short period is so hard that detergents fortified with ammonia or solvents have no effect. These materials can only be removed by physical abrasion, using coarse grit abrasive nylon mesh discs.

The present trend in research is towards polymers with an even greater degree of resistance to detergents. Polymers developed along these lines will have greater durability and will reduce maintenance costs even further.

PREPARATION OF FLOORS FOR WAXING

Correct and thorough preparation of the floor is essential if the best possible results are to be obtained. Where results are poor, usually taking the form of loss of gloss or durability, the fault can nearly always be traced to insufficient initial preparation. Where the floor is correctly prepared, gloss, adhesion, durability and other qualities are shown to their best advantage.

Whilst, in certain circumstances, one type of water emulsion floor wax can be used on top of another type, to obtain the best possible results all previous applications should be removed. If it is intended to apply a water emulsion floor wax to a floor previously treated with a solvent-based wax, it is absolutely essential that all old wax is removed.

Solvent-based wax should be removed using a solvent-based detergent wax remover, described in Chapter 1. Thorough rinsing after using the remover is absolutely essential, as any solvent-based wax remaining on the floor will seriously affect the performance, particularly the gloss, of a water emulsion floor wax.

If a water emulsion floor wax has previously been used it should be removed with an alkaline detergent. It is always advisable to add a little vinegar to the rinsing water to ensure that no alkali

APPLICATION OF WATER EMULSION FLOOR WAX

remains on the floor. After rinsing the floor should be allowed to dry before wax is applied.

Floors which have not previously been waxed should be cleaned with a neutral detergent, using fine or medium grade metal fibre or nylon web pads, if necessary. If the floors are very dirty an alkaline detergent should be used, rinsing thoroughly as above.

New tile floors should not be washed for a period of about 14 days after installation, to enable the adhesive to dry out and harden thoroughly. If after this period it is necessary to clean the floor prior to application of an emulsion floor wax, a weak solution of a neutral detergent in water should be used. The liquid should be applied sparingly; if mopping equipment is used, mops should be just sufficiently damp to clean the floor without an excess of water being present. Alkaline detergents, coarse scouring powders, paste cleaners and solvents, such as paraffin and white spirit, should not be used.

The floor should be allowed to dry before an emulsion floor wax is applied.

It is essential that porous surfaces, such as wood, wood composition, cork and magnesite are sealed before a water emulsion floor wax is applied. Otherwise water could penetrate into these surfaces with possible harm to the floor.

APPLICATION OF WATER EMULSION FLOOR WAX

The wax can be applied using a wide variety of equipment designed for the purpose. If application is to be performed manually, tools such as a sponge mop, ordinary round yarn mop, lambswool applicator or long stranded mop can be used. Long stranded mops are generally used in conjunction with mop buckets and special geared wringers. Emulsion floor wax can also be dispensed from the tank of a polishing/scrubbing machine, if required.

If a sponge mop, ordinary round yarn mop or lambswool applicator is to be used, a small quantity of emulsion should be poured onto the floor and spread evenly with the applicator. Alternatively, the emulsion can be poured into a bucket or shallow tray and the applicator dipped into it before applying the material to the floor.

The operation should be repeated until the whole area of floor is covered. It is important that a thin, even film is applied and that the emulsion floor wax is not allowed to form puddles, particularly if the floor is uneven. If a thick film is applied, the drying time will be considerably extended and the film may be ' spongy ' and soft, with consequent loss of durability.

If a long stranded mop is to be used in conjunction with a mop bucket and geared wringer, the emulsion floor wax is poured into the bucket and the mop immersed in it. The mop should then be placed into the wringer, leaving only the heel above the geared press, and the emulsion wrung from the strands. Thus, when the mop is put on to the floor, the emulsion in the heel is applied and spread evenly with the strands. This method is extremely effective and large areas can be quickly covered.

Once the whole floor area is covered with wax it should be allowed to dry. Under normal conditions this will take about 20–30 min. As the drying operation consists entirely of the evaporation of water, warm, dry conditions with adequate ventilation are ideal. In cold, damp conditions the drying time may be extended.

If a further coat is required it can be applied immediately the preceding coat has dried hard. Traffic can be allowed over the floor as soon as the emulsion floor wax has dried. The longer the final application is left, however, the harder and more durable will be the finish. The top coat should, if possible, be allowed to harden overnight before being subjected to traffic.

Emulsion floor wax is frequently over-applied. Coat upon coat is applied not only in the main traffic lanes but also along the edges of corridors, around the sides of rooms and in other areas not subjected to traffic. Whilst the wax wears and films are kept relatively thin in traffic areas, a 'build-up', consisting of multi-layers of floor wax and dirt, quickly takes place in the remaining areas. This accounts for the unsightly, dirty-brown coloured strips often seen along the edges of well trafficked corridors, particularly in older buildings.

Build-up can, however, be easily prevented by applying wax in traffic areas only, with an occasional coat in non-trafficked areas. If, for example, three coats are to be applied to a freshly cleaned and dewaxed corridor, the first two coats should be applied to the main traffic lane, stopping about 9 in from the wall, the width of a normal floor tile. A rough guide would be to apply floor wax to the centre and leave one tile width untreated along the edges of the corridor. The third coat can then be applied overall to present an even appearance. This is shown diagrammatically in *Figure 4.1*.

This method eliminates the unsightly appearance given by a build-up of emulsion floor wax and also facilitates removal when periodic stripping of all old emulsion from the floor is necessary.

After application, all equipment must be washed in warm soapy water. It is important that cleaning is carried out immediately, otherwise the tough polymers and resins used in water emulsion floor waxes will set hard and make it difficult for the equipment to be used again.

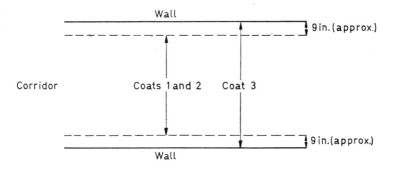

Plan view of a corridor

Figure 4.1. The application of three coats of water emulsion floor wax to a corridor. The first two coats are applied in the main traffic lane, stopping about 9 in, or one tile width, from the wall on either side. The third coat is applied overall

MAINTENANCE OF WAXED FLOORS

Correct maintenance of floors treated with an emulsion wax is essential if they are to be kept in a clean and attractive condition and ensures that maximum durability is obtained from the wax. Perhaps the most important benefit to be gained by maintaining floors correctly is a considerable reduction in the amount of time and effort needed to keep them in a clean and attractive condition, with consequent savings in labour and material costs.

The highest quality materials cannot give optimum results if correct maintenance procedures are not followed. On the other hand, even comparatively poor materials can be made to perform well if used correctly.

The method of maintenance that should be adopted for any particular situation will depend on a number of factors, including the availability of labour, the type of floor and building and the availability and types of floor maintenance machines or mopping equipment.

Whilst many different methods have given, and are still giving satisfactory results, those described below have been found from long experience to suit almost every set of conditions. They are, therefore, offered as a guide and as a starting point, from which variations can be evolved as required.

FLOOR WAXES

Method 1—No Floor Polishing/Scrubbing Machines Available

(1.1) The floor must be in a clean condition before routine maintenance can begin. Any old floor wax should be removed using an alkaline detergent solution. The floor should then be well rinsed with water, to which a little vinegar has been added and, when dry, two coats of a semi-buffable or dry-bright water emulsion floor wax applied. If the floor is worn, a water-based seal may be applied before the floor wax. Alternatively, if a water-based seal is not used, three coats of emulsion floor wax may be required. Each coat should be allowed to dry hard before the next is applied. The final coat should be allowed to harden overnight before traffic is allowed on the floor.

(1.2) Sweep daily to remove surface dust and light soilage. Light traffic marks can be removed by mopping the floor with a solution of 1 cupful of emulsion floor wax/1 gal warm water. A small piece of a nylon web pad, attached to the heel of a long stranded mop, will greatly facilitate the removal of any marks. It is important that the mop is kept damp and that only the minimum amount of solution is applied to the floor.

Stubborn soilage and marks can be removed by the same method, using a solution of a neutral detergent in warm water. Again, it is essential that only the minimum amount of solution is applied to the floor, as excess liquid will soften the surface and impair the durability of the film.

(1.3) Weekly, or at longer intervals as required, apply a further coat to the clean, dry surface. Frequency of application will depend upon the type of water emulsion floor wax used and the nature and volume of traffic.

(1.4) Every 3 months, or longer if applications of emulsion floor wax are made infrequently, strip off all coats of wax using an alkaline detergent solution. Rinse well and repeat the cycle.

Periodic stripping is an integral part of planned maintenance and time for this operation should always be provided in any maintenance programme. If floor maintenance machines are not available, stripping is particularly important to prevent a build-up of emulsion wax taking place. Once a thick build-up occurs it can be very difficult to remove.

Method 2—Floor Polishing/Scrubbing Machines Available; Daily Maintenance Process

(2.1) The floor must be clean before maintenance can begin. Any old wax should be removed using an alkaline detergent solution and the floor well rinsed with water, to which a little vinegar has been added. If the floor is porous or worn a water-based seal should be applied. When dry, apply two or three thin, even coats of a fully buffable or semi-buffable water emulsion floor wax. It is important to avoid over-application as two thin coats will give better durability than thick, spongy coats. For the best possible results each coat should be buffed when dry using a fine grade metal fibre or nylon web pad under a polishing/scrubbing machine. The final coat should be left to harden overnight before traffic is allowed on the floor.

(2.2) Sweep daily to remove surface dust and light soilage. Light traffic marks can be removed by damp mopping the floor daily with a solution of 1 cupful of emulsion floor wax/1 gal warm water. The mop should be just sufficiently damp to leave a thin film of liquid on the surface. When dry, the floor should be buffed using a fine grade metal fibre or nylon web pad, or better, a polishing brush, under a polishing/scrubbing machine.

If marks remain on the floor they can be effectively removed by passing a machine over the area whilst still damp. A fine or medium grade metal fibre or nylon web pad should be used for this purpose. This technique will remove marks, dry and burnish the floor in one operation. Finally, for best possible results, it should be buffed when dry with a polishing brush.

Heavy traffic marks can be removed by the same procedure, using a solution of neutral detergent in water.

(2.3) Weekly, or at longer intervals as required, after cleaning the floor apply a further coat of fully buffable or semi-buffable emulsion floor wax. When dry, buff with a polishing brush under a polishing/scrubbing machine.

(2.4) After 6 or 9 months, depending on the condition of the floor, strip off all old applications of emulsion floor wax using an alkaline detergent solution and repeat the cycle.

Method 3—Floor Polishing/Scrubbing Machine Available; Continuous Maintenance Process

This method is known by various names, including ' dry-cleaning ' ' mist cleaning ' and ' spray cleaning '. The term ' dry cleaning ' is somewhat misleading as some liquid is used. ' Spray cleaning ' is, perhaps, the most appropriate.

(3.1) The floor should be prepared and treated with either a fully buffable or semi-buffable water emulsion floor wax, as described in *Method 2*.

(3.2) Sweep daily to remove surface dust and light soilage. Prepare a solution consisting of approximately 1 cupful of the fully or semi-buffable floor wax/1 gal water. Spray a fine mist over an area of a few square yards and, while still damp, buff the area with a fine grade metal fibre or nylon web pad under a polishing/scrubbing machine.

The action of buffing while still damp will remove any surface soil, dry the floor and buff to a high gloss, all in a few passes of the machine. The next area should then be treated similarly, until the whole floor has been cleaned.

By operating the spray cleaning method the floor undergoing treatment need not be closed to traffic, as only a small area is damp at any one time. Spray cleaning can, therefore, be carried out during normal working hours whilst permitting traffic over the area being cleaned.

(3.3) Weekly, or at longer intervals as required, apply a further coat to the clean, dry surface. Frequency of application will depend upon the type of water emulsion floor wax and the nature and volume of traffic.

(3.4) Every 6 or 9 months, depending on the condition of the floor, strip off all old applications of emulsion floor wax using an alkaline detergent solution and repeat the cycle.

Spray cleaning has been found in practice to be both an efficient and economic method of maintaining floors. It will extend the period between wax stripping operations, keep floors in a clean and attractive condition and is faster than the conventional method of scrubbing and rewaxing.

It is not, however, a cure-all and the method cannot be applied universally. Because a fine mist is applied before buffing, only a little liquid is present. Very dirty floors cannot, therefore, be cleaned by this method and *Method 2* should be adopted. If traffic is light-to-medium and the floors are in a fair-to-good condition, spray cleaning can be used with the advantages described earlier.

To operate spray cleaning satisfactorily a special technique is required. Good results will only be obtained if the method is used properly and not skimped. At least ten or twelve spray cleaning operations are required before optimum resistance to scuff and carbon black heel marks is obtained.

It is essential that only a fine mist is applied. Experience has shown that there is a tendency on the part of operators to apply far too much liquid to the floor. Too much liquid slows the process very considerably and starts a stripping action on the floor which can cause a patchy appearance. Pads also fill quickly, requiring frequent changing or turning over.

The method is designed for speed and does not necessitate the complete removal of every scuff or heel mark with each spray clean. Whilst the majority of marks will be removed during the first machine pass, any more stubborn marks should be given an extra spray. This technique will remove almost all marks and the few, if any, remaining should be left. They will be removed when periodic wax stripping eventually becomes necessary.

A further advantage of the method is that all equipment is carried on the floor polishing/scrubbing machine. Mops, pails, wringers and machine scrubbing equipment are not required, so that corridors and other traffic areas are free and uncluttered. There is no delay, the process is continuous; operators are not kept waiting for washed floors or water emulsion floor wax to dry.

It is not always necessary to spray clean the whole floor. When cleaning corridors, for example, the centre section only should be treated, machine cleaning to about 6 in from the walls. In an office, only the main traffic lanes need be treated regularly, with extra attention given to areas under desks. In areas where traffic is heavy, such as around vending machines, more frequent spray cleaning than is required for normal traffic areas may be necessary.

Even comparatively inexperienced operators can quickly become proficient at the method, which becomes easier with practice and as successive coats become hardened by buffing. Operators quickly learn how often to turn pads over or replace them by looking for the appearance of a fine grey powder on the floor. The powder is caused by a full pad throwing out the dirt it can no longer retain. Used pads should be washed and allowed to dry.

When properly carried out, the spray cleaning method of maintenance will produce floors that are not only clean and attractive, but also safe, with a high degree of slip-resistance.

FACTORS AFFECTING THE CHOICE OF FLOOR WAX

Selection of a floor wax is often carried out in a haphazard and casual manner. As with other floor maintenance materials, discrimination in selection is essential if the best possible results are to be obtained. Whilst a well chosen wax can reduce maintenance costs and present an attractive appearance, a floor can be ruined by use of the wrong type.

Many instances are known of thermoplastic and PVC(vinyl) asbestos tiles being treated with a solvent-based wax, resulting in softened tiles which eventually had to be lifted and the floors relaid.

Before a floor wax is adopted for use in any particular area, it is strongly recommended that a number of factors relating to waxing the floor are considered. Many different types of wax are available and the choice of the best one for a given set of circumstances is often difficult.

A material which has given every satisfaction on one floor may not be the best for use on another, similar floor, because not only do floors vary in their physical composition but also in their relationship to other, outside influences. For example, one floor may be adjacent to an area of bare concrete, so that presence of concrete dust is a constant problem. Another, treated with a water emulsion wax, may be next to a floor treated with a solvent-based wax, resulting in slippery conditions.

Each type of floor wax has its own characteristics, advantages and disadvantages and these should be fully considered and understood before use.

The general properties and methods of maintenance of various types of floor wax have been discussed earlier. Consideration will now be given to the main factors affecting the choice of wax and these are listed below though not in any order of priority. Each should be considered, if only briefly. When all the relevant factors have been considered, the important ones, which most influence the choice of floor wax, will become evident.

Factors affecting the choice of floor wax

    Reason for waxing—the aim.
    Type of floor and whether sealed.
    Location of floor in a building.
    Availability of machines.
    Availability of labour.
    Availability of floor for cleaning.
    Present and future methods of maintenance.
    Durability required.
    Resistance to yellowing on ageing.
    Resistance to slip.
    Flammability.
    Cost, initial and over an extended period.

Each of the above factors is now considered.

FACTORS AFFECTING THE CHOICE OF FLOOR WAX

Reason for Waxing—the Aim

The most important factor that must be considered is the reason for waxing the floor. Most are waxed to improve the appearance and to protect the floor itself from wear, but some are waxed to facilitate maintenance and to reduce maintenance costs.

If protection of the floor itself is the main reason, some thought should also be given to sealing the floor, prior to waxing. It is always advisable to seal a porous floor before waxing.

A few areas are better unwaxed, for example quarry tiles in a kitchen, where spillage of water, grease and fats onto a floor wax may cause slippery conditions. Under these circumstances the floor should be maintained with a solution of detergent in water, rather than by waxing.

A critical examination of the aim may reveal that another method of maintenance would achieve better results. In general, however, particularly where appearance and protection of the floor are concerned, application of a floor wax is essential for good maintenance.

Type of Floor and Whether Sealed

The type of floor under consideration is an extremely important factor. Whether it is sealed must also be taken into account, particularly where porous floors are concerned.

Some floors, for example wood, are harmed by application of a water emulsion floor wax and must only be maintained with a solvent-based wax. Conversely PVC(vinyl) asbestos and thermoplastic tiles, for example, are harmed by solvent and must only be maintained with a water emulsion wax. The type of wax will, therefore, depend upon the floor itself.

If, however, a porous floor is sealed, different circumstances apply. A well sealed wood floor can be satisfactorily maintained with either a solvent-based wax or a water emulsion floor wax because water remains on the surface of the seal and is not in contact with the wood itself.

The type of floor and whether it is sealed or not will greatly influence the decision whether to use a solvent-based wax or a water emulsion floor wax and is a vital factor.

Location of Floor in a Building

The location of a floor is important when considering water emulsion floor waxes. Floors in busy entrance halls and corridors require a material formulated for exceptional durability. Where very dirty conditions are encountered and regular scrubbing with a detergent solution is necessary, a detergent-resistant floor wax should be used.

Under normal conditions an acrylic emulsion with both dry-bright and buffable properties will give excellent results. Where little traffic is encountered a wash-and-wax material could be used with advantage.

In certain circumstances, particularly if the building is very large, it may prove beneficial to use more than one type of water emulsion floor wax. The use of one type in heavily trafficked areas and another in areas with light traffic may show considerable financial savings after only a short period of time.

Availability of Machines

The availability of floor polishing/scrubbing machines is an important factor when considering the type of water emulsion floor wax to use. If there are no machines available for buffing purposes, a material with dry-bright properties should be used; if they are available, either a semi-buffable or fully buffable emulsion floor wax is preferred.

The advantages that can be gained by buffing with machines are many. For example, they save manual effort and considerably greater areas can be treated effectively in the same period of time without operators suffering from fatigue. Also, the gloss of a buffable material can be readily renewed, without the need to apply a further coat. This can be a great advantage over a period of time. If a dry-bright emulsion is applied, the appearance of the floor is at its best immediately after the material has dried and then deteriorates as traffic causes carbon black heel and scuff marks. Sweeping and damp mopping will assist to maintain the appearance of the floor, but after about a week, depending on the nature and volume of traffic, it may have deteriorated to such an extent that a further coat is necessary. If a floor is waxed once weekly, at the end of four weeks six coats will have been applied, assuming two coats initially. This is illustrated diagrammatically in *Figure 4.2*.

Using this method a considerable number will have been applied after only a few months. A build-up of emulsion floor wax can take place and may result in unsightly dark edges along corridors and in rooms, particularly if the material has a tendency to yellow on ageing, and removal is made increasingly more difficult as more coats are applied. If periodic wax stripping is not carried out

FACTORS AFFECTING THE CHOICE OF FLOOR WAX

regularly and frequently, considerable difficulty may be experienced in removing the many layers of emulsion floor wax.

If machines are available and a buffable emulsion is used, the appearance after wear can be renewed by buffing, rather than a further application of floor wax. If buffing is carried out regularly the wax is hardened and durability is improved, as well as the

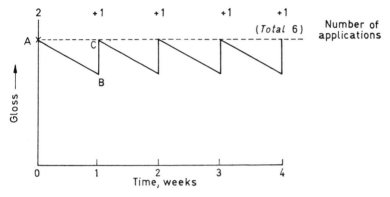

Figure 4.2. Maintenance using a dry-bright emulsion floor wax. Diagrammatic illustration of a floor after two initial coats of dry-bright emulsion floor wax, followed by application once weekly over a period of four weeks. A total of six coats has been applied

A—High initial gloss with two coats dry-bright emulsion floor wax
AB—Deteriorating gloss and appearance over one week due to traffic
B—Unacceptable standard of gloss after one week
BC—Increase in gloss after cleaning and application of one coat of dry-bright emulsion floor wax
C—Renewed gloss after application of dry-bright

appearance. A further coat of wax may only be required after a period of, say, 4 weeks. Many instances are known of buffable emulsion floor waxes remaining in an excellent condition, even under heavy traffic conditions, for a period of 4 months or even longer.

Assuming, however, that a further coat will be required after a period of 4 weeks and that two coats were applied initially, the situation can be illustrated diagrammatically as in *Figure 4.3*.

By using a buffable emulsion floor wax, therefore, considerable savings can be effected with regard to both labour and materials. In addition, the appearance of the floor can be maintained at a high level and the periods between wax stripping operations greatly extended.

FLOOR WAXES

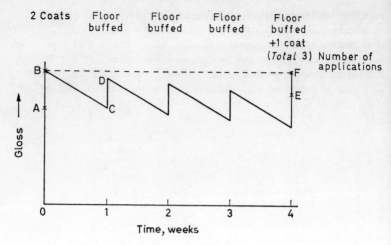

Figure 4.3. Maintenance using a buffable emulsion floor wax. Diagrammatic illustration of a floor after two initial coats of a buffable emulsion floor wax, followed by weekly buffing over a period of four weeks and a further coat at the end of this period. A total of three coats has been applied

A—Initial gloss with two coats buffable emulsion floor wax
B—Gloss after initial buffing.
BC—Deteriorating gloss and appearance over one week due to traffic
C—Unacceptable standard of gloss after one week
CD—Increase in gloss during buffing
D—Renewed gloss level after buffing
E—Unacceptable standard of gloss after buffing
EF—Increase in gloss after one further coat applied and floor buffed
F—Renewed level of gloss after buffing

Availability of Labour

In most organizations, shortage of trained labour is a constant problem. All available labour must, therefore, be effectively employed. In the field of floor maintenance, this can best be carried out by using buffable emulsion floor waxes in conjunction with electric polishing/scrubbing machines.

Buffable emulsion floor waxes generally require less maintenance and therefore less labour, than dry-bright emulsions as they can be spray cleaned, or damp mopped and buffed, whereas floors treated with a dry-bright emulsion require more frequent washing and application of coats.

The use of buffable floor waxes in conjunction with machines and modern floor maintenance equipment is the best method of overcoming problems caused by shortage of labour. On the other hand, if labour is plentiful almost any maintenance programme can be devised to give excellent results.

FACTORS AFFECTING THE CHOICE OF FLOOR WAX

Availability of Floor for Cleaning

Some floors, for example those in office corridors, are readily available for maintenance; others may be in constant use and free for very limited periods only.

The former can be maintained by any of the methods described earlier. If the floor is available for only limited periods, some thought should be given to applying a durable type of floor wax and maintaining by buffing. In this way the amount of maintenance required will be the minimum and results will quickly be obtained in the short time available.

Present and Future Methods of Maintenance

Where maintenance is being effectively carried out it is likely that a suitable type of water emulsion floor wax is already in use. In that event further investigation into the practical aspects of other materials of a similar type may prove to be of advantage.

It should be emphasized that a small sample of water emulsion floor wax applied in an obscure corner will not achieve the desired results. Any trials should be carried out on a large area over a period of months, so that the true merits of the product can be ascertained. Only in this way can durability and removability be properly evaluated.

The future method of maintenance, including a consideration of both labour and machine availability, is a vital factor and influences greatly the type of water emulsion floor wax that should be used.

Durability Required

In general, a wax possessing the longest possible durability is required because the cost of labour is such that stripping and re-application should be carried out as infrequently as possible. Whilst such a wax is essential in entrance halls and heavily trafficked areas, a less durable material may be satisfactory under light traffic conditions, where one that can be readily re-buffed may be preferable to a heavy duty floor wax that cannot be so easily re-buffed.

Resistance to Yellowing on Ageing

This is an important factor, particularly where light coloured floors are concerned. Many floors contain white tiles and any yellowing can quickly ruin the appearance.

Acrylic-based water emulsion floor waxes are, perhaps, the best with regard to resistance to yellowing on ageing. They are extremely light coloured and a floor treated with a material of this kind will retain its colour almost indefinitely.

Resistance to Slip

When floor waxes were in their infancy there was a marked difference in slip resistance properties between various types. It has long been recognized that slip resistance is an extremely important property of any floor wax and is absolutely essential in almost all areas. Because of this fact, the great majority of materials now have good, or even excellent, slip resistance properties.

In general, water emulsion floor waxes have greater slip resistance than solvent-based waxes, largely because of the polymer present in the water-based type. Almost all waxes, however, have a coefficient of friction between leather and floor wax of greater than 0·4, so that the chances of anyone slipping when walking normally are less than one in a million.

As with floor seals, there is no relationship at all between gloss and slip; many high gloss products have extremely good slip resistance properties. Slip resistance is generally improved by regular buffing, so hardening the wax.

Flammability

All solvent-based waxes are flammable and all water emulsion floor waxes are non-flammable. Solvent-based waxes should, therefore, be stored away from naked lights. Whilst they are perfectly safe for all normal uses, if the area in which they are applied is a high fire risk, the use of a water emulsion floor wax should be considered.

Cost, Initial and over an Extended Period

A common fallacy when considering costs is to assume that the cheapest material will be the most economical. This is not often the case. In any cleaning and waxing operation, material costs account for only about 10% of the total, the remaining 90% being that of labour. Greatest consideration, therefore, should be given to keeping the latter at a minimum, which is generally achieved by using the best quality materials available.

A top quality floor wax, although initially dearer, will almost certainly prove to be cheaper over a period of time because of the reduced amount of work and therefore labour costs required to maintain the floor in a clean and attractive condition.

The above are the main factors that should be examined when selecting a floor wax. Once all these have been considered the correct type to use should become evident.

APPENDIX I

FLOOR MAINTENANCE CHART

THE foregoing chapters have dealt in some detail with industrial detergents, floor seals and waxes. From them it will be apparent that discrimination in selection of the materials is essential to success. Each type of floor requires specific methods of treatment and these can vary widely from floor to floor. Whilst some materials will give excellent results, others can do permanent harm and must be avoided. *Table A.1.* is intended to summarize briefly which detergents, seals and waxes should be used and which avoided, on all the main types of floor.

With regard to detergents, the use of neutral and weak alkaline materials is generally quite safe, though on some floors, for example wood, the amount of water should be kept at a minimum. Strong alkaline detergents are also quite safe on most floors when used infrequently for periodic wax stripping operations, but are not, however, generally recommended for routine daily cleaning.

When considering floor seals reference should always be made to the appropriate chapter for details, particularly with regard to the preparation of floors for sealing. The importance of correct preparation cannot be over-emphasized, as the durability of any seal depends to a large extent on its adhesion to the floor.

The floor waxes are rather simpler, as the choice lies between water- and solvent-based materials. It should be remembered that when waxes are applied on top of seals, they do not come into contact with the floor itself and it is the effect only of the wax on the seal that must be considered. Thought should, however, be given to the position that will arise when the seal inevitably starts to wear and no longer forms a protective layer between wax and floor. Reference should be made to Chapter 4, where this aspect is discussed in detail.

Table A.1. *Floor Maintenance Chart. Part 1—Detergents*

Type of Floor	Detergents to Use	Detergents to Avoid
Wood Wood composition Cork	Solvent-based detergents Neutral detergents	Alkaline detergents Abrasive powders Oily materials Detergent crystals
Magnesite	Solvent-based detergents Neutral detergents	Strong alkaline detergents Abrasive powders Oily materials Detergent crystals
Linoleum	Solvent-based detergents Neutral detergents Weak alkaline detergents	Strong alkaline detergents Abrasive powders Oily materials Detergent crystals
Rubber	Neutral detergents Weak alkaline detergents	Strong alkaline detergents Abrasive powders Oily materials Detergent crystals Solvent-based detergents
Thermoplastic tiles PVC(vinyl) asbestos Flexible PVC	Neutral detergents Weak alkaline detergents	Strong alkaline detergents Abrasive powders Oily materials Detergent crystals Solvent-based detergents
Asphalt	Neutral detergents Alkaline detergents Detergent crystals	Oily materials Solvent-based detergents
Concrete	Neutral detergents Alkaline detergents Detergent crystals Solvent-based detergents	Oily materials
Terrazzo Marble	Neutral detergents Mild abrasive powders	Alkaline detergents Detergent crystals Oily materials Solvent-based detergents
Stone	Neutral detergents Alkaline detergents Abrasive powders Detergent crystals Solvent-based detergents	Oily materials

Table A.1. *Floor Maintenance Chart. Part 1—Detergents (cont'd)*

Type of Floor	Detergents to Use	Detergents to Avoid
Quarry tiles	Neutral detergents Alkaline detergents Abrasive powders	Detergent crystals Oily materials Solvent-based detergents

Table A.1. *Floor Maintenance Chart. Part 2—Floor Seals*

Type of Floor	Floor Seals to Use	Floor Seals to Avoid
Wood Wood composition Cork	Solvent-based clear seals	Water-based seals Solvent-based pigmented seals Silicate dressing
Magnesite	Solvent-based clear seals Water-based (coloured) seals	Silicate dressing
Linoleum	Water-based seals (N.B. If necessary a two-pot polyurethane clear seal can be used)	Solvent-based clear seals (Except two-pot polyurethane seals) Solvent-based pigmented seals Silicate dressing
Rubber Thermoplastic tiles PVC(vinyl) asbestos Flexible PVC Terrazzo Marble	Water-based seals	Solvent-based clear and pigmented seals Silicate dressing
Asphalt	Water-based (coloured) seals Two-pot polyurethane clear and pigmented seals Synthetic rubber pigmented seal	Solvent-based clear and pigmented seals (except two-pot polyurethane and synthetic rubber seals) Silicate dressing
Concrete	One- and two-pot polyurethane clear and pigmented seals Synthetic rubber pigmented seal Silicate dressing Water-based seals	Solvent-based clear and pigmented seals (except polyurethane and synthetic rubber seals)

Table A.1. Floor Maintenance Chart. Part 2—Floor Seals (cont'd)

Type of Floor	Floor Seals to Use	Floor Seals to Avoid
Stone	Generally none, but pigmented two-pot polyurethane, synthetic rubber and water-based seals and silicate dressing can be used on some floors	Solvent-based clear seals
Quarry tiles	None	Solvent-based clear and pigmented seals Water-based seals

Table A.1. Floor Maintenance Chart. Part 3—Floor Waxes

Type of Floor	Floor Waxes to Use	Floor Waxes to Avoid
Wood Wood composition Cork	Solvent-based waxes (N.B. If floor is well sealed a water emulsion floor wax can be used)	If unsealed, avoid water emulsion floor wax
Magnesite	Solvent-based waxes (N.B. If floor is sealed a water emulsion floor wax can be used)	If unsealed, avoid water emulsion floor wax
Linoleum Concrete	Solvent-based waxes Water emulsion floor waxes	None
Rubber Thermoplastic tiles PVC(vinyl) asbestos Flexible PVC Asphalt Terrazzo Marble Stone	Water emulsion floor waxes	Solvent-based waxes
Quarry tiles	None (N.B. If a wax is necessary, a water emulsion floor wax can be used)	Solvent-based waxes

APPENDIX II

FAULTS—CAUSES AND REMEDIES

INTRODUCTION

ALL reputable manufacturers of floor maintenance materials, machines and ancillary equipment are very much aware of the need to produce goods which are consistent in quality. To this end, systems of quality control are devised to ensure that all products are manufactured within specified tolerance limits.

Quality control is a very important function in any manufacturing organization. It includes a critical examination of raw materials, work in progress and manufactured materials and a periodic examination of finished goods in stores.

Once the goods have left the manufacturer they may be subject to a wide variety of harmful influences which may affect their performance. For example, in very cold weather a water emulsion floor wax may freeze in transit or in an unheated store room causing the polymer to thicken. Part-used containers of seal may be allowed to stand without lids, causing them to lose solvent and harden.

Incorrect preparation of the surface may also result in an inferior performance. This is particularly the case with respect to floor seals and occasional reports of poor durability can almost always be traced to too little, or incorrect preparation. Floor waxes can also be influenced by this, the most common fault being insufficient rinsing or no rinsing at all.

The vast majority of materials manufactured today give excellent service. Indeed no manufacturer would survive for long if this were not so. The aim of every supplier is to give the best service of which he is capable, both for his customers' satisfaction and for his own.

It is, however, inevitable that snags will arise from time to time. When this occurs, it is helpful to know the likely cause of the fault and perhaps more important, the remedy.

Possible faults, causes and remedies have been listed with respect to industrial detergents, floors seals and floor waxes. The information given in these sections has been compiled from experience gained over a period of very many years. It is intended as a 'First Aid', to help overcome snags that may arise during the course of floor maintenance.

Table A.2. Faults—Causes and Remedies, Part 1. Industrial Detergents

Fault	Possible Cause of Fault	Remedy
(1.1) Loss of colour (or yellowing) of blue and green linoleum after using an alkaline detergent	Effect of alkali on pigment in lino. Some blue and green pigments fade or turn yellow when acted upon by strong alkalis. Tests should be carried out on lino before using an alkaline detergent to ensure pigments are fast to alkalis	If a colour change has taken place, immediately mop with a vinegar solution which may restore the lost colour. If possible, use a neutral detergent rather than an alkaline detergent. If an alkaline detergent is used on blue and green lino, ensure the solution is as dilute as possible. Do not allow an alkaline detergent solution to remain on the floor longer than necessary
(1.2) Excess foam, or difficulty in rinsing foam away	(*a*) Too much detergent being used	Follow label directions. Use as little detergent as necessary
	(*b*) Sponge being used. Due to their nature sponges cause foam	Use another applicator
		Mop with a soap solution as soap destroys foam
(1.3) Lack of foam. (Check that detergent is intended to foam, many are formulated to keep it at a minimum)	(*a*) Insufficient detergent being used	Follow label directions
	(*b*) Detergent solution may be contaminated with soap solution	Empty container, rinse well and remake detergent solution
(1.4) Detergent ineffective for removing wax build-up	(*a*) Wrong detergent being used. (For removing a water emulsion floor wax an alkaline detergent should be used. For removing a solvent-based wax a solvent-based detergent wax remover should be used)	Use the correct type of detergent
	(*b*) Build-up of water emulsion floor wax severe	A number of scrubbings may be required. In very severe cases a fortified detergent (including a solvent) may be necessary in conjunction with an abrasive nylon mesh disc under a scrubbing machine

Table A.2. Faults—Causes and Remedies, Part 1. Industrial Detergents (cont'd)

Fault	Possible Cause of Fault	Remedy
(1.5) Harsh on skin	Continued immersion of hands in any detergent will dry the skin	Instruct users in methods avoiding contact between detergent and skin. If immersion is necessary for any length of time, rubber or PVC gloves should be worn

Table A.2. Faults—Causes and Remedies, Part 2. Floor Seals

Fault	Possible Cause of Fault	Remedy
(2.1) Seal thickened in container	(a) In a one-pot seal, thickening is generally caused by loss of solvent by evaporation	Add a little of the solvent recommended by the manufacturer and stir well. Replace lid
	(b) In a mixed two-pot seal thickening is due to a chemical reaction which takes place between base and accelerator	Reacted material cannot be thinned. Thickened material should be thrown away
(2.2) Slow drying	(a) Floor incorrectly prepared. If wax is present on the floor drying will be delayed	In certain circumstances drying can be speeded up artificially. Because of the wax between the floor surface and seal, however, further trouble, particularly with regard to adhesion and durability, may be experienced later. If incorrect preparation is suspected, clean off the seal from a small area using a nylon web pad and the appropriate solvent. De-wax this small test area again and apply a fresh coat of seal. If satisfactory, treat the whole area in the same way

Table A.2. *Faults—Causes and Remedies, Part 2. Floor Seals (cont'd)*

Fault	Possible Cause of Fault	Remedy
(2.2) Slow drying (cont'd)	(b) Seal applied too thickly. Most one-pot seals dry by air oxidation. If a thick film is applied, the surface will dry but the body of the seal will take longer	Allow time for the seal to dry hard. Good ventilation and warmth will assist
	(c) Low temperature. Cold conditions slow down chemical reactions	Raise temperature by the use of heaters, if available
	(d) Poor ventilation. Solvent must be allowed to evaporate from the seal before the seal can harden	Improve ventilation, if necessary with fans
	(e) Low humidity. Moisture-cured polyurethane seals dry by absorbing moisture from the atmosphere	If humidity is particularly low a special catalyst can be added to the seal
	(f) High humidity. Water-based seals dry by evaporation of water. High humidity will retard evaporation	Apply heat and switch on extractor fans, if available
	(g) A second coat may have been applied before the first was hard dry. If the first coat is not dry, solvent will be trapped beneath the second coat. Escape of solvent through the second coat may take a long time	Allow time to dry thoroughly hard. Each coat should be allowed to dry hard before a further coat is applied
(2.3) Poor adhesion, evidenced by seal: (i) Slow drying, and/or (ii) flaking, and/or (iii) peeling	(a) Floor incorrectly prepared. If all floor wax and dirt have not been removed the seal will not adhere	The only satisfactory remedy for poor adhesion is complete removal of the seal by sanding, or machining with abrasive nylon mesh discs. Then treat a test area to ensure floor is thoroughly clean

Table A.2. *Faults—Causes and Remedies, Part 2. Floor Seals (cont'd)*

Fault	Possible Cause of Fault	Remedy
(2.3) Poor adhesion (*cont'd*)		If only a small area is affected it may be patched up by sanding or machining to remove loose seal and roughen the surface; or when dry, by applying seal to centre and feathering out to edges
	(*b*) Seal incompatible with previous seal on the floor. Some types of seal will not adhere to others	Sand or machine with abrasive nylon mesh discs to remove all old seal, then reapply new seal
	(*c*) Poor preparation when applying new seal to old. With plastic seal particularly, it is essential to roughen the surface of the old seal before applying new	Sand or machine with abrasive nylon mesh discs to remove all old seal, then reapply new seal
	(*d*) Too long a period between coats of seal. In the case of two-pot polyurethane seals, each coat combines chemically with the coat below. Maximum adhesion will only be achieved if coats are applied within a minimum, specified, period	Machine with abrasive nylon mesh discs to remove top coat and to roughen surface. Mop with appropriate solvent. Apply new coat as soon as solvent has evaporated
(2.4) Poor finish. May be: (*i*) streaky (*ii*) patchy (*iii*) rough	(*a*) Floor incorrectly prepared. If all floor wax and dirt are not removed, at the very least the final finish will be poor	Lightly sand or machine with abrasive nylon mesh discs to flatten surface. Mop with appropriate solvent and apply new coat as soon as solvent has evaporated
	(*b*) Dust settling on seal after application. After sanding or machining, all dust must be removed by vacuuming or damp mopping	Lightly sand or machine with abrasive nylon mesh discs to flatten surface. Vacuum or apply mop just sufficiently damp to hold dust and allow to dry. Then apply a further coat of seal

Table A.2. Faults—Causes and Remedies, Part 2. Floor Seals (cont'd)

Fault	Possible Cause of Fault	Remedy
(2.4) Poor finish (*cont'd*)	(*c*) In the case of two-pot polyurethane seals, the presence of water on the floor or in the applicator will cause bubbles to appear in the dried film of seal	Lightly sand or machine with abrasive nylon mesh discs to flatten surface. Vacuum or damp mop to remove bits. Mop with appropriate solvent and apply seal as soon as solvent has evaporated
	(*d*) Too thick an application of seal. The surface may have dried before all solvent has evaporated. Escaping solvent may cause roughness or patchiness	Lightly sand or machine with abrasive nylon mesh discs to flatten surface. Vacuum or damp mop to remove bits. Mop with appropriate solvent and apply seal as soon as solvent has evaporated
	(*e*) Overworking of seal on the floor. Particularly with polyurethane and epoxy seals, overworking with an applicator may cause bubbling, where the applicator is lifted from the floor	Lightly sand or machine with abrasive nylon mesh discs to flatten surface. Vacuum or damp mop to remove bits. Mop with appropriate solvent and apply seal as soon as solvent has evaporated
	(*f*) In the case of water-based seals, the floor may not have been properly rinsed before application of seal	Strip off seal using an approved detergent, rinse well, allow to dry and apply fresh coats of water-based seal. Alternatively, if the area is not too bad, it can be improved by application of a water emulsion floor wax followed by buffing
(2.5) Poor durability (i.e. wearing of seal after a short period). This should not be confused with poor adhesion	(*a*) Inadequate number of coats of seal, taking into account the porosity of the floor and nature of traffic	Thoroughly clean the floor and apply further coats of seal, as necessary
	(*b*) Seal applied too thinly, or thinned with solvent prior to application	Thoroughly clean floor and apply further coats of seal, as necessary

Table A.2. Faults—Causes and Remedies, Part 2. Floor Seals (cont'd)

Fault	Possible Cause of Fault	Remedy
(2.5) Poor durability (cont'd)	(c) Grit carried into the floor from outside	Thoroughly clean the floor and apply further coats of seal, as necessary. Remove source of grit or provide effective mats
	(d) Too frequent scrubbing with coarse grade metal fibre or nylon web pads	Thoroughly clean the floor and apply further coats of seal, as necessary. Check that correct grade pads are used at the correct intervals
	(e) Omission of a finishing coat in a primer/finishing coat system	Thoroughly clean the floor and apply coats of finishing seal, as necessary

Table A.2. Faults—Causes and Remedies, Part 3. Solvent-Based Waxes

Fault	Possible Cause of Fault	Remedy
(3.1) Floor appears slippery	(a) Excess of wax caused by: (i) Build-up over a long period (ii) Too frequent application (iii) Too heavy application	Strip off wax using a solvent-based detergent wax remover. Re-apply a thin coat of wax and buff well
	(b) Outside influences. Oil, solvent or water carried onto the floor by footwear can cause slippery conditions	Remove the outside influences. Use of correct door mats will assist
	(c) Insufficient buffing of wax. Buffing hardens the wax and improves anti-slip properties	Buff the wax regularly, particularly after a new application
	(d) Underfloor heating causing wax to soften	Water emulsion waxes are preferred on floors with underfloor heating

Table A.2. Faults—Causes and Remedies, Part 3. Solvent-based Waxes (cont'd)

Fault	Possible Cause of Fault	Remedy
(3.2) Swirls (machine marks) remain on floor after buffing	(a) Buffing carried out before the floor is dry	Apply a very thin coat of wax, allow to dry thoroughly and buff
	(b) Too heavy application of wax	Machine buff with fine grade metal fibre or nylon web pad. The pad should be changed or cleaned frequently as it picks up excess wax
	(c) Machine brush, or pad, dirty	Clean brush or pad and re-buff
	(d) Incorrect grade of pad used	Fine grade pads only should be used to buff solvent waxes
(3.3) White haze or white marks in wax film	Presence of water either on the floor prior to waxing or applied to the floor before the wax is dry. (The water may be present on the machine used for spreading the wax. If the wax is applied from a machine tank, the tank must be dry)	Strip wax from the floor using a solvent-based detergent wax remover. Apply fresh wax under dry conditions. Check that machine brushes and tank are free from water
(3.4) Wax being walked off in small, paper-thin pieces, which are usually black in colour	Wax build-up. Excess wax must be stripped off from time to time	Strip off wax build-up with a solvent-based detergent wax remover. Reapply fresh wax and buff
(3.5) Slow drying	(a) Inadequate ventilation or low temperature. Solvent waxes dry by evaporation of solvent. Poor ventilation and low temperature will retard drying	Provide adequate ventilation and heat, if possible
	(b) Wax applied too thickly. Wax should be applied in thin, even coats. Two thin coats are better than one thick one	Allow time for wax to dry. Buff with pad to remove excess wax and harden the surface, changing pad regularly

Table A.2. *Faults—Causes and Remedies, Part 4. Water Emulsion Floor Waxes*

Fault	Possible Cause of Fault	Remedy
(4.1) Poor dry-bright qualities. (Check that material is a dry-bright emulsion)	(a) Floor incorrectly prepared. If an alkaline detergent has been used to remove previous wax, all alkaline residue may not have been rinsed away. This residual alkali will attack the fresh coat of emulsion causing, amongst other things, loss of gloss	Strip wax off the floor using an alkaline detergent. Rinse thoroughly, adding a little vinegar to the rinsing water. Apply a fresh coat. Alternatively, if the finish is not bad, buff to harden the film and apply a further coat of wax
	(b) Use of dirty equipment to apply emulsion wax. Detergent remaining in a mop will have an adverse affect on performance when wax is next applied	Keep equipment clean. Strip wax off the floor using an alkaline detergent. Rinse thoroughly, adding a little vinegar to the rinsing water. Apply a fresh coat. Alternatively, if the finish is not bad, buff to harden the film and apply a further coat of wax
	(c) Emulsion wax applied too thinly	Check that coverage is not greater than that given by the manufacturer. Also check that emulsion has not been watered down. Apply a further coat of emulsion wax
	(d) Insufficient number of coats applied. Two or three coats may be necessary to produce a good gloss, particularly on a porous surface	Apply further coats as necessary. On porous surfaces a water-based seal should be used prior to emulsion wax application
	(e) Second coat applied before first was hard dry. The result will resemble that of applying one thick coat (often evidenced by patchy finish and haziness in film)	If appearance is very bad, remedy is to strip wax off the floor using an alkaline detergent. Rinse thoroughly, adding a little vinegar to the rinsing water. Apply a fresh coat. If the

Table A.2. Faults—Causes and Remedies, Part 4. Water Emulsion Floor Waxes (cont'd)

Fault	Possible Cause of Fault	Remedy
(4.1) Poor dry-bright qualities (*cont'd*)		finish is not too bad, buff to harden the film and apply a further coat of wax. If appearance is moderate, buff with a stripping pad under a machine to remove excess wax, followed by application of a fresh coat
(4.2) 'Curdled' appearance in container	(*a*) May be contaminated by inadvertant addition of some incompatible material	Replace with fresh material
	(*b*) May have frozen causing polymer to coagulate	Replace with fresh material
(4.3) Floor appears slippery	(*a*) Emulsion wax application too heavy or too frequent. Excess wax or build-up of wax will cause the surface to become less slip resistant	Strip wax off the floor using an alkaline detergent. Rinse thoroughly adding a little vinegar to the rinsing water. Apply a fresh coat
	(*b*) Insufficient buffing if emulsion wax is buffable type	Machine buff the wax regularly, particularly after a fresh application
	(*c*) Outside influences. Check that oil or solvent is not being carried on to the floor. In wet conditions, traffic will carry water on to the waxed floor with possible danger of slip. Solvent wax may be carried by traffic from an adjoining floor.	Remove outside influences. Prevent entry of oil, solvent or water on to the floor. Correct use of door mats will assist
	A machine brush used for solvent wax must not be used on water emulsion wax	Use separate brushes for areas treated with solvent and water wax
	(*d*) Excess silicone in mops	Check that correct dust mops are being used

Table A.2. Faults—Causes and Remedies, Part 4. Water Emulsion Floor Waxes (cont'd)

Fault	Possible Cause of Fault	Remedy
(4.3) Floor appears slippery (cont'd)	(e) Incorrect quality of floor wax used. For example, application of solvent wax to vinyl tiles may cause slippery conditions	Strip off incorrect grade of wax and apply correct material
(4.4) Poor flow (streaky or patchy finish)	(a) Floor incorrectly prepared. If an alkaline detergent has been used to remove previous wax, all alkaline residue may not have been rinsed away. This residual alkali will attack the fresh coat, causing a streaky and uneven finish, with dull patches	Strip wax off the floor using an alkaline detergent. Rinse thoroughly adding a little vinegar to the rinsing water. Apply a fresh coat. Alternatively, if the appearance is not bad, the finish may be improved by buffing, followed by a fresh coat
	(b) Wax applied to a dirty floor. A dirty floor cannot produce good results and floors should always be cleaned before an emulsion wax is applied	Strip wax off the floor using an alkaline detergent. Rinse thoroughly adding a little vinegar to the rinsing water. Apply a fresh coat
	(c) Contaminated equipment used to apply emulsion wax. Detergents, disinfectants etc. in an applicator can have a damaging effect on an emulsion wax film	Use clean equipment. Strip wax off the floor using an alkaline detergent. Rinse thoroughly adding a little vinegar to the rinsing water. Apply a fresh coat. Alternatively, if the appearance is not bad, the finish may be improved by buffing, followed by a fresh coat
	(d) Emulsion wax applied too thickly. For best results thin coats should be applied, two thin coats being better than one thick one. Check that coverage is not poor, lack of coverage suggests over-application	Strip wax off the floor using an alkaline detergent. Rinse throughly, adding a little vinegar to the rinsing water. Apply a fresh coat. Alternatively, if the appearance is not bad, the finish may be improved by buffing, followed by a fresh coat

Table A.2. Faults—Causes and Remedies, Part 4. Water Emulsion Floor Waxes (cont'd)

Fault	Possible Cause of Fault	Remedy
(4.4) Poor flow (*cont'd*)	(*e*) Second coat applied before first was hard dry	Strip wax off the floor using an alkaline detergent. Rinse thoroughly, adding a little vinegar to the rinsing water. Apply a fresh coat. Alternatively, if the appearance is not bad, the finish may be improved by buffing, followed by a fresh coat
(4.5) Powdering. (A white or light coloured powder appears on the surface of the floor after traffic has been allowed over it. The powder can be swept up leaving no apparent trace on the floor surface)	(*a*) Inadequate preparation. Alkaline detergent left on the floor after wax stripping will react with the fresh coat of emulsion wax and may cause powdering	Strip wax off the floor using an alkaline detergent. Rinse thoroughly, adding a little vinegar to the rinsing water. Apply a fresh coat
	(*b*) Excess wax on floor by applying too heavily or too frequently	Strip wax off the floor using an alkaline detergent. Rinse thoroughly, adding a little vinegar to the rinsing water. Apply a fresh coat
	(*c*) Excessive use of alkaline detergents causing the floor surface to remain in an alkaline state. Alkaline detergents should only be used for periodic stripping. For daily maintenance a neutral detergent, or very weak alkaline detergent solution, should be used	If the floor becomes alkaline it can be neutralized by either washing repeatedly with a vinegar solution, or soaking with a vinegar solution overnight, followed by rinsing with clean water. When the floor is neutral, apply one or two coats of a water-based seal, followed by emulsion floor wax. Use a neutral detergent for normal cleaning
	(*d*) Humid and/or cold conditions. Excessively high humidity and low temperature cause powdering.	Ensure humidity and temperature are not abnormal

Table A.2. Faults—Causes and Remedies, Part 4. Water Emulsion Floor Waxes (cont'd)

Fault	Possible Cause of Fault	Remedy
(4.5) Powdering (cont'd)	There is also evidence that excessively low humidity may cause powdering	
	(e) There is evidence that some sweeping compounds cause powdering, in particular those with a high proportion of calcium chloride. This substance remains on the floor, absorbs moisture and creates artificially humid conditions	Remove dust by damp mopping methods. If sweeping compounds must be used, one with little or no calcium chloride should be chosen
(4.6) Slow drying	(a) Abnormal atmospheric conditions (e.g. high humidity, low temperature) or poor ventilation. Emulsion floor waxes dry by evaporation of water. Conditions must allow evaporation to occur	Adjust temperature and if possible, humidity. Ensure adequate ventilation by opening windows and doors
	(b) Damp floor, caused by lack of good damp course	No remedy other than the provision of a good damp course
(4.7) Poor durability (necessity to apply wax more frequently than seems reasonable). Check that durability is unreasonable under the prevailing traffic conditions	(a) Incorrect detergent or too concentrated a solution of detergent used for daily maintenance	Use correct detergent at appropriate concentration. Where frequent washing with detergent is necessary it is advisable to use a detergent-resistant floor wax
	(b) Incorrect pads being used for buffing or scrubbing. If too coarse a pad is used an emulsion floor wax can be abraded from the floor	Use the finest grade of pad necessary to achieve the desired result
	(c) Abrasive material (grit) carried on to the floor by traffic	Remove abrasive material. Use mats to remove grit from footwear

Table A.2. Faults—Causes and Remedies, Part 4. Water Emulsion Floor Waxes (cont'd)

Fault	Possible Cause of Fault	Remedy
(4.7) Poor durability (cont'd)	(d) Faulty preparation. Alkaline detergent left on the floor prior to waxing will affect the adhesion of a new application of emulsion wax, causing it to be walked off quickly	Strip off wax using an alkaline detergent. Rinse thoroughly, adding a little vinegar to the rinsing water. Apply a fresh coat
	(e) Emulsion wax applied too thinly. Check that coverage is not unusually high	Apply a further coat or coats of emulsion wax as necessary
(4.8) Poor removability	(a) Wrong detergent used for stripping off emulsion wax. Emulsion waxes can only be effectively stripped with alkaline detergents at the correct concentration. Neutral detergents will not act upon wax emulsions	Use the correct detergent at the recommended concentration
	(b) Excessive build-up of wax. Even alkaline detergents at the correct concentration have their limitations	Strip wax build-up from the floor by: (i) Repeated scrubbing with an alkaline detergent (ii) Scrubbing using an alkaline detergent fortified with solvent (iii) Machining with abrasive nylon mesh discs, with an appropriate detergent Check maintenance procedure to prevent excessive build-up in the future
	(c) Incorrect pads being used under a machine	Use the correct, stripping, grade metal fibre or nylon web pad

APPENDIX III

GLOSSARY OF TECHNICAL TERMS*

Abrasive Nylon Mesh Discs These discs are circular and are supplied in a wide range of diameters to fit most makes of floor maintenance machine. They consist of nylon fabric mesh coated with resin-bonded silicon carbide.

Accelerator A substance which increases the speed of a chemical reaction. By common use, the name has become associated with two-pot surface coating materials. The accelerator usually occupies the smaller container and must be added to the larger container, the base, before use.

Acid See *pH*.

Acrylic Resins Manufactured from acrylic acid. They are transparent, water-white and thermoplastic. An acrylic resin conveys the characteristics of toughness, lightness of colour and excellent water resistance.

Alkali See *pH*.

Alkali-soluble Resins A resin soluble in an alkaline solution. Alkali-soluble resins are widely used in water emulsion floor waxes.

Alkyd Varnishes Manufactured from glycerol. They are normally pale in colour and dry rapidly to a glossy, durable film with excellent adhesion. Alkyd varnishes are used in many interior, exterior and stoving paints, and to a lesser extent in floor seals.

All-Resin Emulsion Wax An emulsion wax manufactured entirely from resin constituents. The term applies particularly to those emulsion waxes which consist of a synthetic wax, which could be called a resin, an alkali-soluble resin and a polymer, which is also a resin. The term ' all-resin ' is intended to distinguish this type of material from an emulsion floor wax containing a natural wax, an alkali-soluble resin and a polymer resin.

Base One component of a two-pot surface coating material. It usually occupies the larger container of the two-pot material. It will

* From ' Maintenance of Floors and Floor Coverings ', 7th Edn., Russell Kirby Ltd., 1965.

APPENDIX III

not, by itself, form a film and requires the addition of an accelerator before use.

Buffable Floor Wax This term is given to a floor wax which, when buffed, will give a greater gloss than when not. The waxes can be rebuffed from time to time, as required.

Button Polish A solution of a button lac in an alcohol solvent, usually of the methylated spirit type. Button lac is produced from shellac, the excretion of an insect and is so called because, when the raw shellac is refined, the end product has the appearance of buttons.

Catalyst A chemical substance which is used to accelerate a chemical reaction without itself being permanently changed.

Caustic Destructive or corrosive to living tissue. This term is usually met in connection with caustic soda and caustic potash, two very strong alkaline materials.

Compatability The tolerance of one dissolved substance towards another dissolved substance. For example, in the process of applying a new seal onto an old seal a greater degree of intercoat adhesion is achieved if the two seals are compatible. If the new seal is incompatible with the old seal, special preparation may be necessary to ensure that the new seal will adhere satisfactorily to the old.

Copolymer A very large complex molecule formed by the reaction together of a great number of small molecules of different types. An example is vinyl acetate-acrylate copolymer, a material used in adhesives.

Cure Frequently used in the same sense as the word 'harden'. For example, a film that has fully hardened can be said to be fully cured. The word is usually used in connection with materials hardened by artificial means such as a chemical reaction or stoving at high temperature. It is not used in connection with air-drying materials, for example, oleo-resinous seals.

Detergent A cleansing agent, which may be solvent- or water-based, for removing dirt, etc. It has the advantages over soap in that it is just as effective in hard water as in soft, and does not form scums.

Dilute The verb 'to dilute' means to reduce in strength by addition of water or other appropriate solvent.

Disinfectant Any compound which will destroy micro-organisms. Carbolic acid (phenol) is one of the best known. Recent developments in this field have produced a large number of stronger disinfectants which are both more effective and safer to handle.

GLOSSARY OF TECHNICAL TERMS

Dry-bright Normally refers to a water-based floor wax which, on application, will dry with a glossy appearance. Dry-bright floor waxes are also known as 'self-gloss' emulsion waxes.

Driers Used to accelerate the drying, or hardening process, particularly in air-drying seals.

Drying The process of hardening. Two stages are normally apparent in the drying process:

(a) *Touch dry* The stage at which the film will not mark when pressed lightly with a finger. At this stage the surface has hardened to the extent that it will not retain dust and dirt settling upon it.

(b) *Hard dry* The stage at which the seal or dressing is sufficiently hard to withstand traffic.

Dusting This term is normally applied to concrete floors, and refers to the disintegration of the surface layer of concrete into very fine particles of 'dust'. Almost all concrete floors dust to a greater or lesser extent, depending upon the concrete mix and type and volume of traffic.

Eggshell Finish Subdued gloss of a surface coating material.

Emulsifying Agent A chemical used in the preparation of emulsions to prevent the components from separating. An emulsifying agent is normally only used in small quantities.

Emulsion A very fine suspension of one liquid in another liquid with which it is not miscible. Oil and water are not normally miscible and will separate if blended together. They can, however, be emulsified by the use of emulsifying agents which suspend one liquid in another.

By common use the word has also come to mean the suspension of a wide range of solid materials in water. For example, although wax is a solid, a suspension of wax in water is called water/wax emulsion.

Emulsion Waxes

(a) *Two component systems* A blend of water/wax emulsion and an alkali-soluble resin or shellac. They may or may not dry with a glossy appearance. An increased gloss can be obtained by buffing.

(b) *Three component systems* A blend of a water/wax emulsion, an alkali-soluble resin or shellac and a synthetic polymer resin emulsion. Examples of polymer resins commonly used in the polish industry are polystyrene and acrylates. The water/wax emulsion, alkali-soluble resin and synthetic polymer resin emulsion can be blended in almost any proportions to give emulsion waxes with a wide variety of properties.

Epoxy Resin A synthetic resin manufactured essentially from petroleum derivatives. It is usually supplied in a two-pot form when used in a floor seal. The base component consists of the epoxy resin; the accelerator may be one of a variety of chemicals. In a solvent-free form it is also used for floor laying purposes.

Etching The process of forming small cavities in a surface by the use of a chemical reagent. For example, when sealing concrete floors it is often desirable to etch the surface with an acid. The cavities so formed enable the seal to penetrate further thus ensuring a greater degree of adhesion.

Film A very thin layer of a substance which, in the case of a floor seal, is usually between 5/1000 in and 10/1000 in thick.

Finishing Coat This term is normally applied to a surface coating material used as the top coat of a painting or sealing system usually over a priming coat or undercoat.

Flashing A phenomenon associated with matt paints and seals. It describes the alternate matt and gloss striation effects sometimes left by brushmarks, instead of the uniform matt finish which should be obtained.

Flash Point The temperature at which vapour from a liquid will ignite when exposed to a small flame or spark. The lower the temperature at which ignition takes place, the more flammable is the liquid. For example, acetone, which has a flash point of $-17\cdot8°C$ ($0°F$) will ignite below ordinary room temperature $18\cdot3°C(65°F)$ and is, therefore, very highly flammable; white spirit, on the other hand, has a flash point of $41\cdot1°C(106°F)$, and therefore requires the temperature to be raised before it will ignite.

Freeze–Thaw Stability This property is normally associated with water emulsion floor waxes and water paints and is the resistance of the material to repeated freezing and thawing. One complete freeze–thaw cycle consists of lowering the temperature of the material until it freezes, holding it at that temperature for a specified period and then allowing it to warm to room temperature, when the material again becomes liquid. This can be repeated as required. When a material fails a freeze–thaw stability test, solid ingredients in the emulsion separate from the liquid forming a hard mass. The material is then in an unusable condition. Depending upon the type of emulsion, a material may be completely freeze–thaw stable

GLOSSARY OF TECHNICAL TERMS

over repeated cycles, stable over a limited number of cycles or completely unstable when frozen and thawed once.

Friction The resistance to motion which is called into play when it is attempted to slide one surface over another.

Germicide See *Disinfectant*.

Gloss A shiny surface given by surface coating materials.

Hardener See *Accelerator*.

Intercoat Adhesion The bonding together of two coats, one upon the other, of surface coating materials.

Lacquer The correct definition of a lacquer is ' a solution of film forming substances in volatile solvents '. Drying takes place entirely by evaporation of solvent, leaving the original film-forming substances as a thin film on the surface.

Levelling Also known as ' flow '. Levelling is the property of a surface coating material to flow out and spread itself evenly over the surface, so eliminating applicator or brush marks.

Liquid Wax A combination of wax and solvent, liquid at room temperature.

Matt A smooth, but dull, surface.

Metal Fibre Floor Pads These pads are circular and are supplied in a wide range of sizes to fit most makes of floor maintenance machine. They are generally manufactured in three grades—Coarse, Medium and Fine.

Miscible Two or more liquids are said to be miscible, if, when brought together, they completely intermix to form one liquid. Two or more liquids are said to be immiscible if, when brought together, they will not intermix and separate into two or more layers.

Nylon Web Pads The pads are circular and come in many different sizes to fit most makes of floor maintenance machine. They are generally manufactured in three grades—Coarse, for wax stripping, Medium, for scrubbing and Fine, for buffing.

Oleo-resinous A blend of oil with a resin. The oleo-resinous type is one of the oldest established seals and consists of an oil processed with a resin and combined with solvent and driers. It dries by the action of oxygen in the atmosphere causing the oil and resin to harden. This process is accelerated by the use of driers.

One-pot (*One-pack* or *One-can*) Refers to material packed in a single container and in a ready-for-use condition, without any further modification.

APPENDIX III

Penetrating Seal A seal which will penetrate into the surface on which it is applied. Oleo-resinous seals are penetrating seals, in contrast to some plastic seals which are surface seals and do not penetrate to any great extent.

pH A method of expressing acidity and alkalinity in numerical terms. The pH scale ranges from 0 to 14. 7 is neutral and is the pH of pure distilled water. Materials with a pH below 7 are acidic, the acidity increasing as the pH decreases, materials with a pH above 7 are alkaline, alkalinity increasing as the pH increases (*Figure 1.3*, page 6).

For example, vinegar, a weak acid, has a pH of approximately 3, hydrochloric acid, a strong acid, has a pH of between 0 and 1. Ammonia, a weak alkali, has a pH of approximately 10 to 11, whereas caustic soda, a strong alkali, has a pH of approximately 14. Acids, in general, are harmful to flooring surfaces. Alkalis will not harm floors when used correctly, but floors treated with strong alkalis must always be well rinsed to ensure that all traces are removed.

Phenolic Resin A synthetic resin manufactured basically from phenol. Widely used in many surface coating materials, for example oleo-resinous seals.

Pigment A solid colouring matter which forms a paint when mixed with a suitable liquid. Pigment not only gives the paint its colour, but also its opacity or hiding power.

Plastic A material which will soften when heated. Plastic materials can either be thermoplastic or thermosetting. Thermoplastic materials can be heated and cooled repeatedly without detrimental effect. Thermosetting materials are compositions which undergo chemical change when heated and cannot be reheated without causing damage.

Plastic Seals

(a) *One-pot* The description 'one-pot plastic seal' is commonly given to those seals which do not contain a drying oil and dry by either evaporation of solvent or by a chemical reaction which is activated by evaporation of solvent.

(b) *Two-pot* The description 'two-pot plastic seal' is commonly given to those seals which require the blending together of two components prior to use.

Polymer A very large, complex molecule formed by the reaction together of a great number of small molecules of the same type.

GLOSSARY OF TECHNICAL TERMS

Examples are polystyrene and polyacrylate, materials often used in water-based waxes.

Polystyrene Resin A compound formed by the polymerization of a resin, styrene. In emulsion waxes polystyrene imparts excellent gloss, hardness and levelling.

Polyurethane A polymer formed as the result of a chemical reaction between two types of chemical compounds, namely an isocyanate and a form of polyester. Among many other applications, polyurethanes are used in floor seals and paints. For these purposes they are normally supplied in three different forms:

(a) *Two-pot* The base component is the polyester and the accelerator, or hardener, the isocyanate. The isocyanate is extremely sensitive to water and moisture vapour in the atmosphere and must be protected from them during storage. Once the base and accelerator are mixed, a chemical reaction is started which only stops when the material has solidified.

(b) *One-pot, oil-modified* In these materials the urethane has already been produced and is further combined with an oil or varnish. They are often referred to as 'urethane oils'. Drying takes place by oxidation of the oil or varnish component.

(c) *One-pot, moisture-cured* These materials consist of urethane, already produced, but with an excess of isocyanate present. Once applied, the excess isocyanate attracts water vapour from the atmosphere and hardens the material. The rate of drying will, therefore, depend largely on the humidity, but in this climate there is sufficient moisture in the atmosphere to effect a complete hardening of the film.

Pot-life This term refers to two-pot materials and is the period during which the material is usable once the base and accelerator components have been blended together. After this period, the material will have thickened to such an extent that it cannot be satisfactorily applied.

Priming Coat A priming coat is the first coat applied on previously untreated surfaces. It provides a foundation on which the durability of the finished system largely depends. For example, on wood surfaces the primer is required to be absorbed into the surface in order to obtain a 'key' for subsequent coats. On cement, plaster and concrete surfaces the primer is formulated so that it will resist chemical attack by the alkaline ingredients of the surface on which it is applied.

APPENDIX III

PVC (Polyvinyl chloride) Floor Coverings There are two main types of PVC (Polyvinyl chloride) floor coverings in use today. These are, first, flexible PVC flooring and secondly, PVC(vinyl) asbestos floor tiles. Flexible PVC flooring is supplied in sheet or tile form and both have a smooth wearing surface. PVC(vinyl) asbestos is usually supplied in tile form. They are similar in composition, the main difference being that PVC(vinyl) asbestos contains asbestos fibre, which is not present in flexible PVC floor coverings. Asbestos fibre is adversely affected by many solvents, for example white spirit. Whilst the above are the two main types of PVC floor coverings in use to-day, tiles with intermediate characteristics are produced by varying the relative amounts of polyvinyl chloride and asbestos fibre. It is, therefore, frequently very difficult to distinguish one type of tile from another.

Rafting Rafting is a phenomenon that can occur with sealed new wood block floors, if they have not been properly prepared. It is the movement of a large number of blocks simultaneously, causing a crack to appear in the floor. This can be caused if the blocks are subjected to large changes in moisture content, causing them to shrink excessively, whilst at the same time they are tightly bonded together by means of the seal. Instead of swelling and shrinking individually into existing gaps, the blocks move as a mass causing a large crack to appear.

Reaction Coating A surface coating formed by reaction between two or more chemicals. For example, the film formed by a polyurethane two-pot seal is a 'reaction coating', because a chemical reaction takes place when the base and accelerator are mixed together.

Resin A resin can be either naturally occurring or synthetic and is characterized by being insoluble in water and soluble in a wide range of solvents, for example, white spirit. Naturally occurring resins are adhesive substances obtained from sources such as pine trees. Synthetic resins are made by chemical means. There are many different resins in use in the industry, for example, phenolic and polystyrene resins.

Rivelling This phenomenon can best be described as severe wrinkling. It normally takes place where seal has been applied too thickly and where the surface has dried quicker than the body of the seal, causing the surface to wrinkle.

Seal A floor seal can be described as a permanent or semi-permanent finish which, when applied to a floor, will prevent the entry of dirt and stains, liquids and foreign matter.

GLOSSARY OF TECHNICAL TERMS

Self-gloss See *Dry-bright*.

Shelf-life The period during which a finished product is in a usable condition in its container. After this period the material may be unsuitable for use due to a variety of reasons, for example, thickening in the tin, excessive rusting of the tin, decomposition due to bacterial attack, etc.

Skin A thick layer of material over the surface coating material, for example a paint or floor seal, formed by the oxidation of the surface layer.

Softwood Softwood is wood which belongs to the order Coniferae, or conifers, which includes for example Spruce, Douglas Fir and Longleaf Pitch Pine.

Solid Content The total solid constituents, usually expressed as a percentage, remaining when all solvents are removed from a material.

Solvent Any liquid which will dissolve a solid is a solvent for that solid. Although water is a solvent for many materials, by common use the word 'solvent' has come to mean liquids other than water. White spirit, for example, is a solvent for many resins. Solvent is normally included in a seal to aid application by enabling the material to be spread easily.

Specific Gravity Is the number of times a material is heavier than the same volume of water, at a stated temperature. The weight in pounds of a gallon of material can easily be calculated by multiplying its specific gravity by ten. For example, the specific gravity of white spirit is 0·787. The weight of one gallon of white spirit is, therefore,

$$0.787 \times 10 = 7.87 \, \text{lb}$$

Synthetic Artificial or man-made. Not derived immediately from naturally occuring materials.

Thermoplastic See *Plastic*.

Thinner A liquid added to a paint or varnish to facilitate application. For example, xylene is a thinner widely used in polyurethane seals. Once the seal is applied, the thinner evaporates.

Toxic Poisonous. *Toxicity* is the degree to which a substance is poisonous.

Translucent A material which is translucent will allow light to pass through it, without being transparent.

APPENDIX III

Two-pot (*Two-pack* or *Two-can*) Refers to materials supplied in two separate containers. The contents of one container must be added to the other and the blended material thoroughly mixed before use. The larger container generally contains the base and the smaller the accelerator or 'hardener'.

Urea-formaldehyde A synthetic resin manufactured by heating together two chemicals, urea and formaldehyde. Urea-formaldehyde is widely used in both one- and two-pot seals. The seals cure by the action of an acid catalyst, which, in a two-pot seal, is the accelerator, or hardener, component. They have the characteristic of being almost water-white in colour.

Vinyl Resin A synthetic resin used in the manufacture of many water emulsion paints, floor coverings, etc.

Vinyl Floor Covering See *PVC* (*Polyvinyl chloride*) *Floor Coverings*.

Viscosity Viscosity is the resistance of a liquid to flow; the greater the resistance, the higher is the viscosity. For example, a thick engine oil has a greater viscosity than a thin cycle oil. Viscosity rapidly decreases with increase in temperature.

Wax (*a*) *Natural* A solid material, chemically related to fats. There is a very wide range of naturally occurring waxes. Examples are beeswax, a soft wax, produced from the sugar of food eaten by bees, formed as a secretion in the bee's stomach, and carnauba wax, a hard wax, produced from the leaves of trees found mainly in Brazil.

(*b*) *Synthetic* There is also a very wide range of synthetic waxes. A well known example of a soft wax is paraffin wax, derived from petroleum. Polyethylene is an example of a harder synthetic wax frequently used in both water- and solvent-based polishes.

Wetting Agent A wetting agent is used to reduce the surface tension between a solid and a liquid. In detergents, a wetting agent is included to loosen dirt from the surface to which it is attached.

White Spirit A solvent derived from the distillation of petroleum and generally known as 'turpentine substitute'. It is widely used in the polish, paint and varnish industries.

APPENDIX IV

CONVERSION TABLES

MEASURES OF WEIGHT—AVOIRDUPOIS AND COMMERCIAL

1 ounce (oz)	=	16 drams (dr)
1 pound (lb)	=	16 oz
1 stone (st.)	=	14 lb
1 quarter (qr)	=	2 st. = 28 lb
1 hundredweight (cwt)	=	4 qr = 112 lb
1 ton	=	20 cwt = 2240 lb

METRIC UNITS OF WEIGHT

1 kilogramme (kg)	=	1000 grammes (g)
1 quintal (q)	=	100 kg
1 tonne	=	1000 kg

EQUIVALENT UNITS OF WEIGHT

Avoirdupois		*Metric*
1 ounce	=	28·35 g
1 pound	=	453·6 g = 0·454 kg
1 stone	=	6·35 kg
1 quarter	=	12·7 kg
1 hundredweight	=	50·802 kg
1 ton (2240 lb)	=	1016·04 kg = 1·016 tonnes

Metric		*Avoirdupois*
1 gramme	=	0·035 oz
1 kilogramme	=	2·2046 lb = 35·274 oz
1 quintal	=	1·968 cwt = 220·462 lb
1 tonne	=	0·984 tons

APPENDIX IV
U.S.A. UNITS

The fundamental units are the metre and the gramme, but the pound and the yard are also in general use. A U.S. ton, or 'short' ton, of 2000 lb is widely used.

U.S.A.	*British*		*Metric*
1 short ton	= 0·8928 tons	= 2000 lb	= 907·184 kg
1 short cwt	= 0·8928 cwt	= 100 lb	= 45·359 kg
1·12 short tons	= 1 ton	= 2240 lb	= 1016 kg = 1·016 tonnes
1·12 short cwt	= 1 cwt	= 112 lb	= 50·802 kg
1·1023 short tons	= 0·9842 tons	= 19·69 cwt	= 1 tonne = 1000 kg

LENGTH

1 statute mile = 1760 yards (yd) = 5280 feet (ft) = 63,360 inches (in)

1 yard = 0·9144 metres (m) = 91·44 centimetres (cm)

1 kilometre (km) = 1000 (m) = 100,000 cm

1 metre = 1·0936 yd = 39·37 in = 3·2808 ft

AREA

1 square yard (yd^2) = 9 ft^2 = 0·836 m^2

1 square foot (ft^2) = 144 in^2 = 929 cm^2

1 square metre (m^2) = 1·196 yd^2 = 10·764 ft^2

CAPACITY

1 cubic yard (yd^3) = 27 ft^3 = 0·7646 m^3

1 cubic metre (m^3) = 35·315 ft^3 = 1·30795 yd^3

1 Imperial gallon (Imp gal) = 4 quarts = 8 pints = 4·546 litres

1 litre (l) = 1·76 pints = 0·22 Imp gal = 1000 millilitres (ml)

1 pint = 0·568 l = 0·125 Imp gal

CONVERSION TABLES

U.S.A. Gallons

1 U.S. gal	=	0·8327 Imp gal	=	3·785 l
1 Imp gal	=	1·2 U.S. gal	=	4·546 l
1 litre	=	0·264 U.S. gal	=	0·22 Imp gal

THERMOMETRICAL

Fahrenheit: Freezing Point = 32°
Boiling Point = 212°
Centigrade: Freezing Point = 0°
Boiling Point = 100°

To convert Fahrenheit to Centigrade—

$$(°F - 32) \times \frac{5}{9} = °C$$

e.g. To convert 68°F to Centigrade—

$$(68 - 32) \times \frac{5}{9} = \frac{36 \times 5}{9} = 20°C$$

and to convert Centigrade to Fahrenheit—

$$\left(°C \times \frac{9}{5}\right) + 32 = °F$$

e.g. To convert 20°C to Fahrenheit—

$$\left(\frac{20 \times 9}{5}\right) + 32 = 36 + 32 = 68°F$$

OTHER USEFUL FACTORS

¼ Imp gal	=	40 fluid oz
1 Imp pint	=	20 fluid oz
1 tablespoonful	=	½ fluid oz
1 teaspoonful	=	⅛ fluid oz
1 litre	=	35·2 fluid oz
1 fluid oz	=	28·4 ml

1 gal water (at 62°F) weighs 10 lb
1 litre water (at 62°F) weighs 2·2 lb

INDEX

Accelerator, 46
Acid cleaners, 21
Acid-sensitive
  emulsion floor wax, 133
  polymers, 126
Acrylic polymer resins, 67, 124
Additives, 126
Adhesion of floor seals, 88
Alkaline
  degreasers, 22
  detergents, 16, 17
  fortified detergents, 19
  powder detergents, 19
Alkali-soluble resin, 117, 122
Ammonia, 125, 135
Ammonia-sensitive floor wax, 134
Amphoteric detergents, 15
Anionic detergents, 11, 13
Application of
  oleo-resinous seal, 34
  polyurethane
    1-pot seal, 43
    2-pot seal, clear, 53
    2-pot seal, pigmented, 63
  seal, 101
  silicate dressing, 71
  synthetic rubber seal, 59
  urea-formaldehyde
    1-pot seal, 39
    2-pot seal, 49
  water-based seal, 68
  water emulsion floor wax, 137
Asphalt, 94
  removal of oil from, 23
  use on, of
    oleo-resinous seal, 34
    polyurethane
      1-pot seal, 42
      2-pot seal, clear, 52
      2-pot seal, pigmented, 62
    solvent wax, 108
    synthetic rubber seal, 58
    urea-formaldehyde
      1-pot seal, 38
      2-pot seal, 49

Asphalt—*contd.*
  use on, of—*contd.*
    water-based seal, 67
    water emulsion floor wax, 116, 118

Base, 46
Beeswax, 106, 107, 109
Benzalkonium chloride, 14
Bleach, 21
Bronze, 21
Brushes, 101
  use with
    oleo-resinous seal, 34
    polyurethane
      1-pot seal, 43
      2-pot seal, clear, 53
      2-pot seal, pigmented, 63
    synthetic rubber seal, 59
    urea-formaldehyde
      1-pot seal, 39
      2-pot seal, 49
Brush cleaning solvent, for
  oleo-resinous seal, 35
  polyurethane
    1-pot seal, 43
    2-pot seal, clear, 54
    2-pot seal, pigmented, 64
  silicate dressing, 72
  synthetic rubber seal, 59
  urea-formaldehyde
    1-pot seal, 39
    2-pot seal, 50
  water-based seal, 69
Build-up, emulsion floor wax, 19, 23, 138
Button polish, 27, 75, 90

Candelilla wax, 110
Carnauba wax, 110, 121
Cationic detergent, 14
Caustic soda, 20
Ceresin wax, 111

INDEX

Chemical resistance of
  oleo-resinous seal, 36
  polyurethane
    1-pot seal, 45
    2-pot seal, clear, 56
    2-pot seal, pigmented, 65
  silicate dressing, 73
  synthetic rubber seal, 61
  urea-formaldehyde
    1-pot seal, 41
    2-pot seal, 51
  water-based seal, 70
Citric acid, 21
Clean Air Act, 1
Coalescing agents, 127
Colour of
  oleo-resinous seal, 34
  polyurethane
    1-pot seal, 43
    2-pot seal, clear, 53
    2-pot seal, pigmented, 63
  silicate dressing, 71
  synthetic rubber seal, 59
  urea-formaldehyde
    1-pot seal, 39
    2-pot seal, 49
  water-based seal, 68
Compatability of seals, 89, 100
Composition of
  oleo-resinous seal, 33
  polyurethane
    1-pot seal, 42
    2-pot seal, clear, 52
    2-pot seal, pigmented, 62
  silicate dressing, 71
  synthetic rubber seal, 58
  urea-formaldehyde
    1-pot seal, 38
    2-pot seal, 48
  water-based seal, 67
Concrete, 94
  removal of oil from, 23, 24
  use on, of
    oleo-resinous seal, 33
    polyurethane
      1-pot seal, 42
      2-pot seal, clear, 52
      2-pot seal, pigmented, 62
    silicate dressing, 71, 72
    solvent wax, 108
    synthetic rubber seal, 58

Concrete—*contd.*
  use on, of—*contd.*
    urea-formaldehyde
      1-pot seal, 38
      2-pot seal, 49
    water-based seal, 68
    water emulsion floor wax, 118
Copolymers, 117, 125
Cork, 89
  use on, of
    oleo-resinous seal, 33
    polyurethane
      1-pot seal, 42
      2-pot seal, clear, 52
    silicate dressing, 71
    solvent wax, 107
    urea-formaldehyde
      1-pot seal, 38
      2-pot seal, 48
    water-based seal, 68
    water emulsion floor wax, 116, 118
Cotton wax, 110

Detergents
  acid, 21
  acidity of, 6
  action, as wetting agent, 4
  alkaline, 16
  alkalinity of, 6
  amphoteric, 15
  anionic, 11, 13
  cationic, 14
  caustic, 20
  crystals, 22
  definition of, 3
  hydrophilic component, 4
  hydrophobic component, 4
  neutral, 11
  non-ionic, 11
  requirements of, 3
  selection of, 8
  solvent wax removers, 24
  surface tension, 4
  types, 8
Detergent-resistant emulsion floor wax, 134
Dirt,
  definition of, 1
  emulsification, 5
  grease, 3
  nature of, 3

INDEX

Dirt—*contd.*
  organic matter, 3
  particulate, 3
  removal of, 4
  rinsing, 6
  suspension, 5
Dry-bright emulsion floor wax, 128
Dry-cleaning, 141
Drying time of
  oleo-resinous seal, 35
  polyurethane
    1-pot seal, 44
    2-pot seal, clear, 55
    2-pot seal, pigmented, 64
  silicate dressing, 72
  synthetic rubber seal, 60
  urea-formaldehyde
    1-pot seal, 40
    2-pot seal, 50
  water-based seal, 69
Durability of seals, 105
  oleo-resinous seal, 36
  polyurethane
    1-pot seal, 45
    2-pot seal, clear, 56
    2-pot seal, pigmented, 65
  silicate dressing, 72
  synthetic rubber seal, 60
  urea-formaldehyde
    1-pot seal, 40
    2-pot seal, 51
  water-based seal, 69
Dust allaying oil, 27

Emulsion, 116
Epoxy
  esters, 74
  2-pot seal, 75
Esparto wax, 110
Etching crystals, 94

Fischer-Tropsch wax, 122
Flammability of seals, 46
Flexible PVC, 96
  use on, of
    oleo-resinous seal, 34
    polyurethane
      1-pot seal, 42
      2-pot seal, clear, 52
      2-pot seal, pigmented, 62

Flexible PVC—*contd.*
  use on, of—*contd.*
    solvent wax, 108
    synthetic rubber seal, 58
    urea-formaldehyde
      1-pot seal, 38
      2-pot seal, 49
    water-based seal, 67
    water emulsion floor wax, 116, 118
Floors, treatment with
  oleo-resinous seal, 33
  polyurethane
    1-pot seal, 42
    2-pot seal, clear, 52
    2-pot seal, pigmented, 62
  silicate dressing, 71
  synthetic rubber seal, 58
  urea-formaldehyde
    1-pot seal, 38
    2-pot seal, 48
  water-based seal, 67
Floor oil, 27
Floor seals, 26
  definitions of, 28
  effect of stiletto heels on, 30
  history of, 26
  non-slip properties, 31
  requirements of, 29
  types of, 32
Floor wax, factors affecting choice of, 143–150
Fluorochemicals, 127
Fully buffable emulsion floor wax, 128
Furniture polish, 112

Ghedda wax, 110
Grease, removal of, 22, 24
Gymnasium oil, 27

Hardener, 46
Hardening time of
  oleo-resinous seal, 36
  polyurethane
    1-pot seal, 44
    2-pot seal, clear, 55
    2-pot seal, pigmented, 64
  silicate dressing, 72
  synthetic rubber seal, 60

INDEX

Hardening time of—*contd.*
  urea-formaldehyde
    1-pot seal, 40
    2-pot seal, 50
    water-based seal, 69
Hexachlorophene, 9
High-solids emulsion floor wax, 130
Hydrochloric acid, 20–22

Indicator papers, 18
Interval between coats of
  oleo-resinous seal, 35
  polyurethane
    1-pot seal, 44
    2-pot seal, clear, 55
    2-pot seal, pigmented, 64
  silicate dressing, 72
  synthetic rubber seal, 60
  urea-formaldehyde
    1-pot seal, 40
    2-pot seal, 50
  water-based seal, 69

Lambswool applicator, 102, 137
  for
    oleo-resinous seal, 34
    polyurethane
      1-pot seal, 43
      2-pot seal, clear, 53
      2-pot seal, pigmented, 63
    synthetic rubber seal, 59
    urea-formaldehyde
      1-pot seal, 39
      2-pot seal, 49
    water-based seal, 68
Lavatory bowl cleaner, 21
Lime encrustations, 21
Linoleum, 91, 96
  effect of alkaline detergent, 17
  use on, of
    oleo-resinous seal, 33, 34
    polyurethane
      1-pot seal, 42
      2-pot seal, clear, 53
      2-pot seal, pigmented, 62
    solvent wax, 108
    synthetic rubber seal, 58
    urea-formaldehyde
      1-pot seal, 38
      2-pot seal, 49
    water-based seal, 67

Linoleum—*contd.*
  use on, of—*contd.*
    water emulsion floor wax, 116, 118
Liquid wax, 112, 114
  removal of, 24, 25

Magnesite, 93
  use on, of
    oleo-resinous seal, 33
    polyurethane
      1-pot seal, 42
      2-pot seal, clear, 52
      2-pot seal, pigmented, 62
    solvent wax, 108
    synthetic rubber seal, 58
    urea-formaldehyde
      1-pot seal, 38
      2-pot seal, 48
    water-based seal, 68
    water emulsion floor wax, 116, 118
Maintenance of
  oleo-resinous seal, 36
  polyurethane
    1-pot seal, 44
    2-pot seal, clear, 55
    2-pot seal, pigmented, 64
  silicate dressing, 72
  synthetic rubber seal, 60
  urea-formaldehyde
    1-pot seal, 40
    2-pot seal, 50
  water-based seal, 69
  waxed floors, 139–143
Marble, 98
  use on, of
    oleo-resinous seal, 34
    polyurethane
      1-pot seal, 42
      2-pot seal, clear, 52
      2-pot seal, pigmented, 62
    silicate dressing, 71
    solvent wax, 108
    synthetic rubber seal, 58
    urea-formaldehyde
      1-pot seal, 38
      2-pot seal, 49
    water-based seal, 67
    water emulsion floor wax, 118
Metal-complexed polymer, 125
Microcrystalline wax, 121
Milkstone, 21

INDEX

Mist-cleaning, 141
Modified nitrocellulose, 74
Mohair roller applicator, 102
Moisture-cured polyurethane, 41
Montan wax, 110, 121
Mops, 34, 68, 101, 137, 143

Nekal A, 2
Neutral detergent, 11
Non-ionic detergent, 11

Odour of
  oleo-resinous seal, 35
  polyurethane
    1-pot seal, 43
    2-pot seal, clear, 54
    2-pot seal, pigmented, 63
  silicate dressing, 71
  synthetic rubber seal, 59
  urea-formaldehyde
    1-pot seal, 39
    2-pot seal, 50
  water-based seal, 68
Offices, Shops and Railway Premises Act, 1
Oil removal, 22, 24
Oil-modified polyurethane, 41
Oleo-resinous seals, 32–37
One-pot
  plastic seals, 37
  polyurethane seals, 41–46
  urea-formaldehyde seals, 37–41
Ouricury wax, 110
Oxalic acid, 21
Ozokerite wax, 111

Palm wax, 110
Paraffin, 25
Paraffin wax, 110
Paste wax, 112
  removal of, 24, 25
Penetrating seals, 33
pH scale, 6
  values, 7, 11, 12, 14, 15, 16, 18, 20
Phosphoric acid, 21
Pigmented seals, 57
Plastic seals,
  1-pot, 37
  2-pot, 46

Plasticizers, 126
  benzylbutyl phthalate, 126
  dibutyl phthalate, 126
  tributoxyethyl phosphate, 126
  tricresyl phosphate, 126
Polish, 106
Polyesters, 2-pot, 74
Polyethylene wax, 111, 122
Polymer resins, 117, 123
  acid sensitive, 126
  acrylic, 67, 124
  copolymers, 117, 125
  metal-complexed, 125, 134
  polystyrene, 117, 124
Polystyrene polymer resin, 117, 124
Polyurethane seals,
  1-pot, 41–46
  2-pot, clear, 52–57
  2-pot, pigmented, 62–66
Public Health Acts, 1
PVC(vinyl) asbestos, 96
  use on, of
    oleo-resinous seal, 34
    polyurethane
      1-pot seal, 42
      2-pot seal, clear, 52
      2-pot seal, pigmented, 62
    solvent wax, 108
    synthetic rubber seal, 58
    urea-formaldehyde
      1-pot seal, 38
      2-pot seal, 49
    water-based seal, 67
    water emulsion floor wax, 116, 118

Quarry tiles, 99
  removal of cement from, 22
  use on, of
    oleo-resinous seal, 34
    polyurethane
      1-pot seal, 42
      2-pot seal, clear, 52
      2-pot seal, pigmented, 62
    silicate dressing, 71
    solvent wax, 108
    synthetic rubber seal, 58
    urea-formaldehyde
      1-pot seal, 38
      2-pot seal, 49
    water emulsion floor wax, 118
Quaternary ammonium compounds, 14

INDEX

Rafting, 89
Recoatability of
  oleo-resinous seal, 36
  polyurethane
    1-pot seal, 45
    2-pot seal, clear, 56
    2-pot seal, pigmented, 65
  silicate dressing, 73
  synthetic rubber seal, 61
  urea-formaldehyde
    1-pot seal, 41
    2-pot seal, 51
  water-based seal, 70
Removability of
  oleo-resinous seal, 37
  polyurethane
    1-pot seal, 46
    2-pot seal, clear, 56
    2-pot seal, pigmented, 65
  silicate dressing, 73
  synthetic rubber seal, 61
  urea-formaldehyde
    1-pot seal, 41
    2-pot seal, 51
  water-based seal, 70
Re-sealing, 99
Rinsing of alkaline detergent, 16, 18
Roller applicators,
  for seal, 102
    oleo-resinous seal, 34
    polyurethane
      1-pot seal, 43
      2-pot seal, clear, 53
      2-pot seal, pigmented, 63
    synthetic rubber seal, 59
    urea-formaldehyde
      1-pot seal, 39
      2-pot seal, 49
Rubber, 97
  use on, of
    oleo-resinous seal, 34
    polyurethane
      1-pot seal, 42
      2-pot seal, clear, 52
      2-pot seal, pigmented, 62
    solvent wax, 108
    synthetic rubber seal, 58
    urea-formaldehyde
      1-pot seal, 38
      2-pot seal, 49
    water-based seal, 67
    water emulsion floor wax, 116, 118

Sanitizing compounds, 14
Seals,
  comparison tables, 84–87
  factors affecting choice of, 76–84
  water-based, 66
Semi-buffable emulsion floor wax, 128
Shelf life of
  oleo-resinous seal, 34
  polyurethane
    1-pot seal, 42
    2-pot seal, clear, 53
    2-pot seal, pigmented, 63
  silicate dressing, 71
  synthetic rubber seal, 58
  urea-formaldehyde
    1-pot seal, 38
    2-pot seal, 49
  water-based seal, 68
Shellac, 117, 122
Shellac lacquers, 75
Shellac wax, 110
Silicate dressings, 70–73
Silicone, 112
Slip resistance, 118
Soap, 8
  consumption of, 2
  formula, 9
  history, 2
  properties, 9
  scum, 2
  surface tension, 2
Soapless detergents, 10
Sodium metasilicate, 22
Sodium tripolyphosphate, 16, 19
Solvents, 111
Solvent-based detergent wax removers, 24
Solvent wax, 106
  removal of, 24, 25
Splintering, 119
Spray-cleaning, 141
Stainless steel, 21
Stiletto heels, 30
Stone, 99
  use on, of
    oleo-resinous seal, 34
    polyurethane
      1-pot seal, 42
      2-pot seal, clear, 52
      2-pot seal, pigmented, 62
    silicate dressing, 71
    synthetic rubber seal, 58

INDEX

Stone—*contd.*
  use on, of—*contd.*
    urea-formaldehyde
      1-pot seal, 38
      2-pot seal, 49
    water-based seal, 68
Styrenated alkyds, 74
Styrene-acrylate copolymers, 117, 125
Styrene/Butadiene lacquers, 75
Sulphurous acid, 21
Surface tension, 2, 4
Surfactants, 2, 3
Synthetic rubber seal, 57–61

Terrazzo, 98
  removal of cement from, 22
  use on, of
    oleo-resinous seal, 34
    polyurethane
      1-pot seal, 42
      2-pot seal, clear, 52
      2-pot seal, pigmented, 62
    silicate dressing, 71
    solvent wax, 108
    synthetic rubber seal, 58
    urea-formaldehyde
      1-pot seal, 38
      2-pot seal, 49
    water-based seal, 67
    water emulsion floor wax, 116, 118
Thermoplastic tiles, 96
  use on, of
    oleo-resinous seal, 34
    polyurethane
      1-pot seal, 42
      2-pot seal, clear, 52
      2-pot seal, pigmented, 62
    solvent wax, 108
    synthetic rubber seal, 58
    urea-formaldehyde
      1-pot seal, 38
      2-pot seal, 49
    water-based seal, 67
    water emulsion floor wax, 116, 118
Three component systems, 117
Tung oil, 33
Turk's head brushes, 59, 63, 103
Turpentine, 111
Two component systems, 117

Two-pot
  epoxy resins, 75
  plastic seals, 46
  polyesters, 74
  polyurethane seals, clear, 52–57
  polyurethane seals, pigmented, 62–66
  urea-formaldehyde seals, 48–52

Urea-formaldehyde seals,
  1-pot, 37–41
  2-pot, 48–52
Urethane-oil, 41

Ventilation, 101, 138
Vinegar, 18, 96, 140, 141

Wash-and-wax emulsion floor waxes, 131
Water-based seals, 66–70
Water emulsion floor waxes, 115
  acid-sensitive, 133
  ammonia sensitive, 134
  application, 137
  coloured, 118
  detergent-resistant, 134
  dry-bright, 128
  fully buffable, 128
  high-solids, 130
  semi-buffable, 128
  stripping, 17
  types of, 127
  wash-and-wax, 131
Watering can, 71
Waxes, 109–111, 120–122
  Beeswax, 106, 107, 109
  Candelilla, 110
  Carnauba, 110, 121
  Ceresin, 111
  Cotton, 110
  Esparto, 110
  Fischer-Tropsch, 122
  Ghedda, 110
  Liquid, 112, 114
  Microcrystalline, 121
  Montan, 110, 121
  Ouricury, 110
  Ozokerite, 111
  Palm, 110
  Paraffin, 110

INDEX

Waxes—*contd.*
 Paste, 112
 Polyethylene, 111, 122
 Shellac, 110
 Solvent-based, 106
 Water emulsion, 115
Wetting agents, 126
White spirit, 25, 111, 112
Wood, 89
 use on, of
  oleo-resinous seal, 33
  polyurethane
   1-pot seal, 42
   2-pot seal, clear, 52
  silicate dressing, 71
  solvent wax, 107
  urea-formaldehyde
   1-pot seal, 38
   2-pot seal, 48
  water-based seal, 68

Wood—*contd.*
 use on, of—*contd.*
  water emulsion floor wax, 116, 118, 119
Wood composition, 89
 use on, of
  oleo-resinous seal, 33
  polyurethane
   1-pot seal, 42
   2-pot seal, clear, 52
  silicate dressing, 71
  solvent wax, 107
  urea-formaldehyde
   1-pot seal, 38
   2-pot seal, 48
  water-based seal, 68
  water emulsion floor wax, 116, 118, 119

Zinc, 125, 135
Zirconium, 125, 135